Entropy Principle for the Development of Complex Biotic Systems

Entropy Principle for the Development of Complex Biotic Systems

Organisms, Ecosystems, the Earth

Ichiro Aoki
(Formerly) Department of Systems Engineering
Faculty of Engineering
Shizuoka University
Japan

ELSEVIER

AMSTERDAM • BOSTON • HEIDELBERG • LONDON • NEW YORK • OXFORD
PARIS • SAN DIEGO • SAN FRANCISCO • SINGAPORE • SYDNEY • TOKYO

Elsevier
32 Jamestown Road London NW1 7BY
225 Wyman Street, Waltham, MA 02451, USA

First edition 2012

British Library Cataloguing-in-Publication Data

A catalogue record for this book is available from the British Library

Library of Congress Cataloging-in-Publication Data

A catalog record for this book is available from the Library of Congress

ISBN: 978-0-323-28232-1

This book has been manufactured using Print On Demand technology. Each copy is produced to order and is limited to black ink. The online version of this book will show color figures where appropriate.

Contents

Preface

The knell of the bells at the Gion temple
Echoes the impermanence of all things.
The colour of the flowers on its double-trunked tree
Reveals the truth that to flourish is to fall.
He who is proud is not so for long,
Like a passing dream on a night in spring.
He who is brave is finally destroyed,
To be no more than dust before the wind.

The Heike Story[1]

Understanding the nature of the development of complex systems is a significant subject in recent science and will be in the future. Although many kinds of investigation on complexity—concepts of self-organization, self-adaptation, edge of chaos, self-organized criticality, and so on—are going forward, these problems are tough, and progress (if) is slow.

However, this line of research is not the only one dealing with complexity, and other approaches may be possible. The nature of complex systems can be investigated from nonreductionistic, phenomenological, and holistic viewpoints. A possibility is to use thermodynamics, specifically the concept of entropy production (a measure of activity), to study macroscopic biological objects. Such studies are presented here. The results are concise, solid, and definitive with classical beauty, in contrast to some other reductionistic nonlinear analyses of complex systems.

The properties of complex macroscopic systems with a biological function are examined here. However, the natures of abiotic and biotic complexity may be different, and the properties of microscopic and macroscopic complexity may also differ.

The time-course of entropy production for biotic systems, from their beginning to their end, whether individual organisms or ecological systems, turns out here to be of a gross diphase nature (with fluctuations): an initial rapid increase and a later slow decrease. This tendency may be universally applied to any system with definite structure and function.

Consequently, it can be proposed in more detail that natural living systems evolve rapidly in the initial stage of their lives to an ordered and solid state, and then they age gradually to a final state of death (chaos). This proposal is different

[1] Japanese classic in thirteenth century, the author(s) is/are not specified. The story in the context of the Buddhism philosophy of impermanence (translated by P. G. O'Neill).

from the well-known hypothesis of Kauffman (1995, 2000) that "complex adaptive systems evolve to the edge of chaos." From that statement questions arise: What happens after the edge of chaos for macro-living systems? Is the edge of chaos the end of biological evolution? And so on.

In each hierarchy of science, there are entirely new laws and concepts (Anderson, 1972). The laws of complexity may differ from abiotic to biotic objects, or from microscopic to macroscopic matters, or from partial views to whole views, or from reductionistic to nonreductionistic approaches, and so on. The arguments presented here are based on biotic, macroscopic, view-from-whole, and nonreductionistic standpoints.

Acknowledgment

The draft is partially preedited by my daughter Keiko.

Ichiro Aoki
Ashiya-City, Japan
2009

1 Thermodynamics and Living Systems

1.1 Thermodynamics

Complex systems, such as individual organisms and ecological systems, are macroscopic, and thermodynamics is a branch of science that is adequate for such macroscopic objects.

At the end of nineteenth century, classical physics had been established, and it was supported by three main props: classical mechanics (Newton), electromagnetic theory (Maxwell–Faraday), and thermodynamics (Joule–Helmholtz–Mayer–Clausius–Lord Kelvin–Boltzmann). It was considered at that time that no other new physical principles remained and that only applications of the three fields remained for future studies in physics. However, in the twentieth century, the revolution in physics is well known, and two of the three principles (Newtonian mechanics and Maxwell–Faraday's electromagnetic theory) have been changed to completely new paradigms (i.e., quantum versions). Only thermodynamics is not so fragile, has not changed, and has been valid until recently (Bennett, 1987; Rubí, 2008; Serreli, Lee, Kay, & Leigh, 2007), and it may perhaps remain so in the future.

Thermodynamics is robust because, in addition to its very solid empirical bases and its dealing with rather homogeneous materials, it is more or less nonreductionistic and systems theoretical ("holological" according to Hutchinson, 1964). This means that the general scheme of systems does not necessarily fully depend on the detailed properties of constituent subsystems and sub-subsystems and their interactions. Nowadays, thermodynamics is not necessarily as popular in physics as in the past, but it is a useful tool in chemical thermodynamics, biological thermodynamics, engineering thermodynamics, and so on.

This systems theoretical characteristic is appropriate for dealing with macroscopic complex systems as a whole from nonreductionistic viewpoints, without the equivocal and nondefinitive results inherent in many prevalent studies, such as nonlinear physiology (Shelhamer, 2007).

In the ensuing chapters, complex biotic systems are investigated by the use of the principles of thermodynamics. Thermodynamics consists of two laws: the First and the Second Laws. The First Law is the law of energy. The energy concept has

Entropy Principle for the Development of Complex Biotic Systems. DOI: 10.1016/B978-0-12-391493-4.00001-9

been extensively employed and well known in the natural, social sciences, and even in our daily lives. In the biological sciences, bioenergetics, energy budget, biocalorimetry, and ecological energetics, among others, are examples of studies using the energy concept. The First Law is discussed in the appendix as it relates to the energy budgets of living systems and is not described here.

1.1.1 Entropy[1] Law

Little has been known about the implications of entropy, especially in biology, although entropy is as significant a concept as energy in thermodynamics: the Second Law is the law of entropy. Hereafter, entropy concepts in biological systems are the main themes of the discussion in this monograph.

Entropy is a rather difficult concept in the textbooks on physics. Entropy is a measure of the quality of energy: the higher the entropy is, the lower the quality of energy becomes. Also, it is a measure of the randomness or disorder in microscopic structures, according to Boltzmann in the nineteenth century, as is well known. However, it should be noted that entropy is *not a universal measure of order or disorder* (Berry, 1995).

For an isolated system, the Second Law states that the change of the system's entropy content in irreversible processes [ΔS(irrev)] is always larger than in reversible processes [ΔS(rev)] and that the latter is zero (the classical Principle of Clausius, 1865)

$$\Delta S(\mathrm{irrev}) > \Delta S(\mathrm{rev}) = 0 \tag{1.1}$$

Since biological objects are not isolated systems, Eq. (1.1) cannot be applied to biology.

Biological systems are open systems that exchange energy and matter with their surroundings. For open systems, the change of entropy content of a system (ΔS) is the sum of two terms: entropy flow ($\Delta_e S$) and entropy production ($\Delta_i S$). Entropy flow is the entropy that is brought into or out of the system and that is associated with flows of energy and matter. Entropy production is the entropy that is produced by irreversible processes occurring within the system. The Second Law for open systems asserts that entropy production in irreversible processes [$\Delta_i S$(irrev)] is always larger than in reversible processes [$\Delta_i S$(rev)] and that the latter is zero (Nicolis & Prigogine, 1977)

$$\Delta_i S(\mathrm{irrev}) > \Delta_i S(\mathrm{rev}) = 0 \tag{1.2}$$

Thus, the Second Law for open systems is formulated in terms of entropy production.

[1] Do not confuse it with entropy in information theory (Shannon, 1948). This name is only the result of a joke by von Neumann and Shannon; thermodynamical entropy has nothing to do with so-called information entropy (Müller, 2007).

In open systems near thermodynamic equilibrium, entropy production always decreases with time and approaches a minimum stationary level according to the minimum entropy production principle (Nicolis & Prigogine, 1977)

$$\frac{d}{dt}\Delta_i S(\text{irrev}) < 0 \tag{1.3}$$

However, biological systems are not near thermodynamic equilibrium but are far from equilibrium; hence Eq. (1.3) cannot be applied to biological systems.[2]

What entropy principle, if any, is applicable to living systems that are open and far from equilibrium? The following chapters show the answer.

1.1.2 Entropy Production: Quantifiability in Living Systems

Thermodynamical variables are divided into two classes: state variables and process variables. With regard to the entropy concept, the state variable is entropy content, and the process variables are entropy flow and entropy production.

It is possible to imagine the state variable (the entropy content) of biological systems *in a purely conceptual context.* Biological organisms are in states that are far from equilibrium; however, even in such conditions, the entropy content of organisms can be considered to exist. Prigogine argued that the entropy of a nonequilibrium state can be defined if the materials are dense and the variations of densities with space and time are small; these conditions are fulfilled in biological systems (Nicolis & Prigogine, 1977). Also, Landsberg (1972) asserted that, for a class of nonequilibrium states, extensive variables such as entropy content exist, and he called this statement the Fourth Law of Thermodynamics. It will be evident that, according to Prigogine, common biological systems are in states for which extensive variable entropy exists and that the Fourth Law holds.

However, it is quite questionable whether entropy content can be measured for living systems. The measurement of the absolute value of the entropy content of an organism calls for measurements of heat capacity from the absolute temperature 0 K to ordinary temperatures and of latent heat at phase transition, if any, for the organism.[3] Measured at an extremely low temperature, the organism is in a "dead" state: its internal organized structure and function are completely destroyed. The entropy content of organisms thus measured have no significant meanings for organisms that must have the essential features of life—specific organization and biotic function. An alternative method for measuring entropy content is to measure

[2] The strong influence of the doctrine of Prigogine and Wiame (1946) based on Eq. (1.3), which had dominated many biothermodynamicists for a long time, had already been denied in developmental biology (Zotin & Zotina, 1993), as well as for macrobiotic systems, as shown later.

[3] The Nernst–Planck heat theorem (the so-called Third Law of Thermodynamics) asserts that entropy content at 0 K is zero for complete crystal solids; hence absolute values of entropy could be measured. However, it turns out that entropy at 0 K is not necessarily zero for some other materials, such as glasses and perhaps biotic systems.

at higher reference temperatures (not at 0 K but near to ordinary temperatures), in which organisms can exist in living state. However, the choice of an adequate reference temperature causes a problem. Reference temperature differs for various kinds of organisms because the temperature dependence of biochemical and physiological conditions differs among organisms.[4] Hence, the relative entropy content thus measured may not be useful for the comparative study of various kinds of organisms. It should be noted, in biological science, that the comparative study is one of the most powerful tools for revealing the fundamental characteristics of biological systems, and it has been recognized from the time of Darwin to the present day of modern molecular biology and biochemistry.

Also, the calculation of the entropy content of living systems is an extremely formidable task (it is really impossible), even for the simplest organisms, such as bacteria. The difficulty is easily understood by recognizing that even a single bacterium contains various kinds of organized structures with tremendous variety and numbers of molecules undergoing intricate and specific biochemical reactions. Even a simple organism differs vastly in complexity from, say, an ideal homogeneous gas confined in a cylinder under constant temperature and pressure, for which the entropy content can easily be calculated, as is done in the textbooks.[5] Hence, it is impossible to develop quantitative thermodynamical considerations based on the measured or calculated entropy content of living systems.

However, entropy flow and entropy production—process variables—can quantitatively be estimated by means of physical methods and calculations from the *observed* energetic data of biological objects. Methods of calculation are described in later chapters. Thus, it is possible to develop entropic considerations based on numerical values of entropy flow and entropy production.

Thus, entropy production is a significant quantity due to the property of Eq. (1.2) and due to the quantifiability of living systems. Entropy is produced anywhere at any time when processes are irreversible. If the irreversibility of a process is higher, more entropy is produced. Hence, entropy production is a measure of the strength of the irreversibility of processes. Since almost all motions and reactions actually occurring in nature are irreversible (except some carefully controlled laboratory experiments), entropy production is also a measure of the magnitude of the activity of natural processes. This activity consists of physical activity (the strength of the process of energy flow and the transportation of matter), of chemical activity (the strength of the chemical reaction), and of biological (including human) activity (the strength of the biological interaction).

The following chapters show the methods of calculation and the results of entropy flow and entropy production mainly for plant leaves, animals, humans,

[4] Consider homeotherms and poikilotherms, as well as animals with hibernation and nonhibernation.

[5] Incidentally, it should be noted that living systems such as the simplest organisms (bacteria) differ greatly in their degree of complexity from simple systems such as ideal gases, Bénard cells, Belousov–Zhabotinsky reactions, networks of lightbulbs, sandpiles, and so on. More importantly, there is the essential difference in the presence and absence of biological function.

ecological systems, and the Earth. They also clarify the general trends of the time course of entropy production in the development of biotic systems.

1.1.3 Maxwell's Demon

Entropy productions of plants, animals, humans, ecological systems, and the Earth, as calculated from observed energetic data, are shown (in later chapters) to be all positive, and hence the Second Law of Thermodynamics certainly holds [Eq. (1.2)] for these systems, against some old assertions (Beier, 1962; Trincher, 1967; cited from Zotin, 1978). Thus, Maxwell's demon (Maxwell, 1871), an agent presumed to violate the Second Law, does not exist for these systems. This is also shown for computational physics (Bennett, 1987), for the recent experiment of a nanoscale machine (Serreli et al., 2007), for some small systems (Rubí, 2008), and for macro-biotic systems to be shown in the following chapters. These results also show that these macrosystems cannot be perpetual-motion machines of the second kind because the Second Law demonstrates the impossibility of making such machines (Principle of Ostwald).[6]

Net entropy flows into biotic systems turn out to be negative; that is, biotic systems absorb negative entropy from the environment. This is the physical basis for maintaining organized structure and function in biotic systems, in the sense of Schrödinger (1944).

[6] The Second Law of Thermodynamics is expressed by the principles of Thomson, Clausius, Ostwald, and Planck. For other versions of the Second Law, refer to Lewis (2007) and Beattie and Oppenheim (1979). They are all physically equivalent, though they are not described here.

2 Plant Leaves

From the energy budget of a broad-plant leaf in moderate conditions, the entropy fluxes into or out of the leaf due to solar radiation, infrared (IR) radiation, evaporation of water, and heat conduction are calculated. Net entropy flow into the leaf is negative. On the assumption that the entropy in the leaf is in a steady state, the entropy production in the typical broad leaf in moderate conditions (the solar energy absorbed by both sides of the leaf is $E_{solar} = 0.0602\ J\ cm^{-2}\ s^{-1}$) becomes $S_{prod} = 1.8 \times 10^{-4}\ J\ cm^{-2}\ s^{-1}\ K^{-1}$. The positiveness of the entropy production shows that the Second Law of Thermodynamics certainly holds in the plant leaf (Aoki, 1987a; Chapter 1). Entropy productions in other conditions and other species (Aoki, 1989b) are also calculated. The entropy production in the leaf (S_{prod}) becomes a linear function of the solar energy absorbed by the leaf E_{solar} : $S_{prod} \simeq (30.6\ E_{solar}) \times 10^{-4}$.

A theorem is presented: the entropy production in plant leaves oscillates during the period of 1 day, paralleling the daily solar energy absorbed by leaves. The most entropy production in plant leaves will be due to the physical scattering and absorption of incident solar radiation by various components and particles in leaves, followed by the subsequent conversion of solar radiation energy to heat energy; these processes cause entropy production. The contribution of photosynthesis to entropy production is shown to be very small because photosynthesis uses only about 1% or less of solar radiation energy incident on plant leaves.

2.1 Broad Leaves (Aoki, 1987a)

Entropy flows and entropy productions in broad-plant leaves are calculated, based on the energetics given by Nobel (1970) and by Gates (1963), by means of the theoretical methods developed by Aoki (1987b, 1998). Then a theorem on entropy production in leaves is presented: entropy production in plant leaves oscillates in steady state with the period of 1 day, paralleling the daily solar energy absorbed by the leaves.

For an understanding of the logical context of this chapter or later ones, some fundamental knowledge of the physics of radiation entropy is required. This is described, for example, in Planck (1959, 1988), Landsberg (1961), Spanner (1964), Landsberg and Tonge (1979), and Aoki (1982a, 1982b, 1983, 1987b, 1998) (also Section 7.1). Some nonphysics readers may skip Sections 2.1 and 2.2 and go to Sections 2.3 and 2.4.

Entropy Principle for the Development of Complex Biotic Systems. DOI: 10.1016/B978-0-12-391493-4.00002-0

2.1.1 Entropies Associated with Shortwave Solar Radiation

Consider a horizontal, flat, and thin broad leaf exposed to full solar radiation at sea level where the energy flux of global (total) radiation is $E_i = 1.20\ \mathrm{cal\ cm^{-2}\ min^{-1}} = 0.0837\ \mathrm{J\ cm^{-2}\ s^{-1}}$ (see the top line of table 7.3 of Nobel, 1970). Assume that the ratio of the energy of diffuse solar radiation to that of global radiation is $d = 0.20$. Also suppose that the reflectivity of solar radiation by the earth's surface is $r = 0.20$ and that the absorptivity of the leaf to solar radiation is $\alpha = 0.60$ (table 7.3 of Nobel, 1970).

Let us calculate solar radiation entropies absorbed by the leaf surface, following a line of discussion described in the appendix of Aoki (1987b) or Section 7.1. First, consider direct solar radiation. The energy of direct solar radiation incident per unit time on a unit area of the leaf surface is

$$E_{dr} = E_i \times (1 - d) = 0.0670\ \mathrm{J\ cm^{-2}\ s^{-1}} \tag{2.1}$$

The corresponding solar entropy flux is given by [see Eq. (A10) of Aoki (1987b) or Eq. (7.9) in Section 7.1]

$$\begin{aligned} S_{dr} &= 2.31 \times 10^{-4}\ \mathrm{K^{-1}} \times E_{dr} \\ &= 0.155 \times 10^{-4}\ \mathrm{J\ cm^{-2}\ s^{-1}\ K^{-1}} \end{aligned} \tag{2.2}$$

Next, the energy of diffuse (scattered) solar radiation incident on the leaf is

$$\begin{aligned} E_{sc} &= E_i \times d \\ &= 0.0167\ \mathrm{J\ cm^{-2}\ s^{-1}} \end{aligned} \tag{2.3}$$

The specific intensity of diffuse solar radiation is given by [see Eq. (B2) of Aoki (1987b) or Eq. (7.11)]

$$K_1 = \frac{E_{sc}}{\pi} = 5.33 \times 10^{-3}\ \mathrm{J\ cm^{-2}\ s^{-1}} \tag{2.4}$$

On the other hand, the specific intensity of extraterrestrial solar radiation incident on the Earth is [Eq. (B3) of Aoki (1987b) or Eq. (7.12)]

$$K_0 = 1.99 \times 10^{3}\ \mathrm{J\ cm^{-2}\ s^{-1}} \tag{2.5}$$

Then, the "emissivity" of diffuse solar radiation is expressed as [Eq. (B4) of Aoki (1987b) or Eq. (7.13)]

$$\varepsilon = \frac{K_1}{K_0} = 2.68 \times 10^{-6} \tag{2.6}$$

and the function $X(\varepsilon)$ becomes

$$X(\varepsilon = 2.68 \times 10^{-6}) = 4.52 \tag{2.7}$$

where [Eqs (B6) and (B7) of Aoki (1987b) or Eqs (7.15) and (7.16)]

$$X(\varepsilon) = \frac{45}{4\pi^4}\frac{1}{\varepsilon}\int_0^{\infty} y^2[(x+1)\ln(x+1) - x\ln x]\,\mathrm{d}y \tag{2.8}$$

$$x = \frac{\varepsilon}{e^y - 1} \tag{2.9}$$

which is introduced by Landsberg and Tong (1979). By means of these values and the temperature of the Sun, $T_0 = 5{,}760$ K, the entropy of diffuse solar radiation incident per unit time on a unit area of the leaf surface is obtained as [Eq. (B10) of Aoki (1987b) or Eq. (7.19)]

$$\begin{aligned} S_{sc} &= \frac{4}{3}\frac{E_{sc}}{T_0}X(\varepsilon) \\ &= 0.175 \times 10^{-4}\ \mathrm{J\ cm^{-2}\ s^{-1}\ K^{-1}} \end{aligned} \tag{2.10}$$

Next, the energy of reflected solar radiation by the Earth's surface incident on the leaf is

$$\begin{aligned} E_{rf} &= E_i \times r \\ &= 0.0167\ \mathrm{J\ cm^{-2}\ s^{-1}} \end{aligned} \tag{2.11}$$

The specific intensity of reflected solar radiation is given by [see Eq. (C2) of Aoki (1987b) or Eq. (7.21)]

$$\begin{aligned} K_1 &= \frac{E_{rf}}{\pi} \\ &= 5.33 \times 10^{-3}\ \mathrm{J\ cm^{-2}\ s^{-1}} \end{aligned} \tag{2.12}$$

The "emissivity" of reflected solar radiation is expressed as [Eq. (C3) of Aoki (1987b) or Eq. (7.22)]

$$\varepsilon = \frac{K_1}{K_0} = 2.68 \times 10^{-6} \tag{2.13}$$

where K_0 is given by Eq. (2.5). Then, the function $X(\varepsilon)$ becomes

$$X(\varepsilon = 2.68 \times 10^{-6}) = 4.52 \tag{2.14}$$

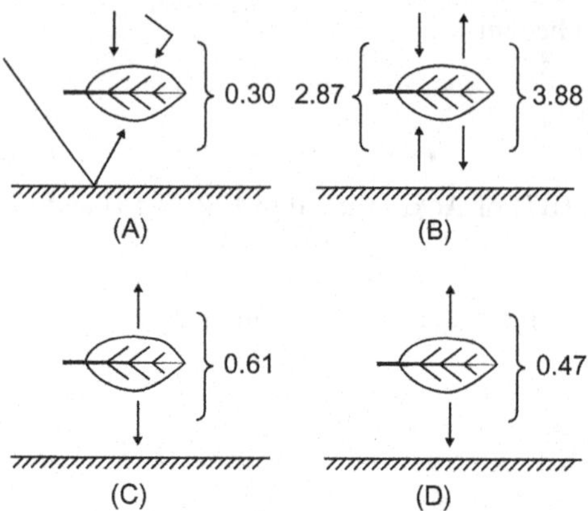

Figure 2.1 Entropy flows into or out of a horizontal broad leaf in units of 10^{-4} J cm^{-2} s^{-1} K^{-1} = J m^{-2} s^{-1} K^{-1}. (A) Direct, diffuse, and reflected solar entropy flows absorbed by the leaf. (B) IR entropy flows absorbed and emitted by the leaf. (C) Entropy flow associated with the evaporation of water. (D) Entropy flow associated with heat conduction.

where $X(\varepsilon)$ is given by Eqs (2.8) and (2.9). The entropy of reflected solar radiation incident per unit time on a unit area of the leaf surface is written as [see Eq. (C7) of Aoki (1987b) or Eq. (7.26)]

$$\begin{aligned} S_{\rm rf} &= \frac{4}{3}\frac{E_{\rm rf}}{T_0}X(\varepsilon) \\ &= 0.175 \times 10^{-4}\ \mathrm{J\ cm^{-2}\ s^{-1}\ K^{-1}} \end{aligned} \tag{2.15}$$

Incidentally, $S_{\rm rf}$ equals $S_{\rm sc}$. This is because the value $d = 0.20$ is, by chance, the same as the value $r = 0.20$ in this case.

The total shortwave entropy flux incident on the leaf is

$$S_{\rm dr} + S_{\rm sc} + S_{\rm rf} = 0.505 \times 10^{-4}\ \mathrm{J\ cm^{-2}\ s^{-1}\ K^{-1}} \tag{2.16}$$

Since the absorptivity of the leaf to solar radiation is $\alpha = 0.60$, the shortwave solar entropy absorbed by the leaf is given by

$$(S_{\rm dr} + S_{\rm sc} + S_{\rm rf}) \times \alpha = 0.303 \times 10^{-4}\ \mathrm{J\ cm^{-2}\ s^{-1}\ K^{-1}} \tag{2.17}$$

The result is shown in Figure 2.1A.

2.1.2 Entropies Associated with IR Radiation

First, consider the IR radiation from the sky. The sky can be regarded as a black body of an effective temperature $T_{\rm sky} = -30°\mathrm{C} = 243$ K (the top line of table 7.3

of Nobel, 1970). The entropy flux emitted by the black body of a temperature T is given by $S = (4/3)\sigma T^3$, where $\sigma = 5.67 \times 10^{-12}$ J cm^{-2} s^{-1} K^{-4} is the Stefan–Boltzmann constant (Planck, 1959, 1988). Hence, the entropy flux emitted by the sky to the leaf is obtained as

$$\begin{aligned} S_{\text{sky}} &= \frac{4}{3}\sigma(T_{\text{sky}})^3 \\ &= 1.085 \times 10^{-4}\ \text{J cm}^{-2}\ \text{s}^{-1}\ \text{K}^{-1} \end{aligned} \tag{2.18}$$

Similarly, since the effective temperature of the ground is $T_{\text{grd}} = 20°\text{C} = 293$ K (the top line of table 7.3 of Nobel, 1970), the entropy flux emitted by the ground to the leaf is given by

$$\begin{aligned} S_{\text{grd}} &= \frac{4}{3}\sigma(T_{\text{grd}})^3 \\ &= 1.902 \times 10^{-4}\ \text{J cm}^{-2}\ \text{s}^{-1}\ \text{K}^{-1} \end{aligned} \tag{2.19}$$

The total IR entropy flux from the sky and the ground is

$$S_{\text{sky}} + S_{\text{grd}} = 2.986 \times 10^{-4}\ \text{J cm}^{-2}\ \text{s}^{-1}\ \text{K}^{-1} \tag{2.20}$$

Since the absorptivity of the leaf to IR radiation is $\alpha_{\text{IR}} = 0.96$ (table 7.3 of Nobel, 1970), the IR entropy absorbed by the leaf becomes

$$(S_{\text{sky}} + S_{\text{grd}}) \times \alpha_{\text{IR}} = 2.867 \times 10^{-4}\ \text{J cm}^{-2}\ \text{s}^{-1}\ \text{K}^{-1} \tag{2.21}$$

The leaf itself also emits IR radiation to the surroundings. The temperature of the leaf is, in this case, $T_{\text{lf}} = 25°\text{C} = 298$ K (the top line of table 7.3 of Nobel, 1970). The IR emissivity of the leaf is $\varepsilon_{\text{IR}} = 0.96$ (table 7.3 of Nobel, 1970), and the function $X(\varepsilon)$ becomes

$$X(\varepsilon_{\text{IR}} = 0.96) = 1.01 \tag{2.22}$$

where $X(\varepsilon)$ is given by Eqs (2.8) and (2.9). In general, the specific intensity of entropy radiation emitted by the gray body of the temperature T and of the emissivity ε is given by [Eq. (6.3) of Aoki (1982a)]

$$L = \frac{1}{\pi}\frac{4}{3}\varepsilon\sigma T^3 X(\varepsilon) \tag{2.23}$$

and the entropy flux emitted by the gray body is

$$S = \int_0^{2\pi} \mathrm{d}\phi \int_0^{\pi/2} L\cos\theta\,\sin\theta\,\mathrm{d}\theta = \pi L \tag{2.24}$$

Hence, the IR entropy flux emitted by the leaf is written as

$$\begin{aligned} S_{\mathrm{lf}} &= 2 \times \frac{4}{3} \varepsilon_{\mathrm{IR}} \sigma (T_{\mathrm{lf}})^3 X(\varepsilon_{\mathrm{IR}}) \\ &= 3.880 \times 10^{-4}\ \mathrm{J\ cm^{-2}\ s^{-1}\ K^{-1}} \end{aligned} \tag{2.25}$$

where the factor 2 is necessary because IR radiation is emitted by both sides of the leaf. These results are shown in Figure 2.1B.

2.1.3 Entropy Associated with Heat Flow due to Evaporation of Water and Conduction

The heat loss of the leaf due to the evaporation of water during transpiration is $Q_{\mathrm{evp}} = 0.26\ \mathrm{cal\ cm^{-2}\ min^{-1}} = 0.0181\ \mathrm{J\ cm^{-2}\ s^{-1}}$ for the leaf of a typical mesophyte (Nobel, 1970, p. 359). The temperature of the leaf is $T_{\mathrm{lf}} = 25°\mathrm{C} = 298\ \mathrm{K}$, as already shown (top line of table 7.3 of Nobel, 1970). Therefore, the entropy flux out of the leaf due to transpiration is

$$S_{\mathrm{evp}} = \frac{Q_{\mathrm{evp}}}{T_{\mathrm{lf}}} = 0.609 \times 10^{-4}\ \mathrm{J\ cm^{-2}\ s^{-1}\ K^{-1}} \tag{2.26}$$

Likewise, since the heat loss of the leaf due to heat conduction is $Q_{\mathrm{cnd}} = 0.20\ \mathrm{cal\ cm^{-2}\ min^{-1}} = 0.0140\ \mathrm{J\ cm^{-2}\ s^{-1}}$ (Nobel, 1970, p. 359), the entropy flux out of the leaf due to heat conduction is

$$S_{\mathrm{cnd}} = \frac{Q_{\mathrm{cnd}}}{T_{\mathrm{lf}}} = 0.468 \times 10^{-4}\ \mathrm{J\ cm^{-2}\ s^{-1}\ K^{-1}} \tag{2.27}$$

These results are shown in Figure 2.1C and D.

2.1.4 Entropy Flow due to Mass-Flow Associated with Photosynthesis

The entropy contents of matters in the standard condition (298.15 K, 1 atm) engaged in photosynthesis are as follows:

CO_2 (gas) 213.64 J K^{-1} mole^{-1},
H_2O (liq.) 69.94 J K^{-1} mole^{-1},
O_2 (gas) 205.03 J K^{-1} mole^{-1},
$C_6H_{12}O_6$ (aq.) 264.01 J K^{-1} mole^{-1}.

Hence, the entropy change per 1 mole of photosynthesized glucose is

$$-[6 \times (213.64 + 69.94) - 6 \times (205.03) - 264.01] = -207.29\ \mathrm{J\ K^{-1}}$$

Or, the entropy change per 1 mole of CO_2 fixed by photosynthesis is

$$\frac{-207.29}{6} = -34.55\ \mathrm{J\ K^{-1}\ mole^{-1}}$$

Since a typical net rate of CO_2 fixation by photosynthesis is 1.1 (nmole cm^{-2} s^{-1}) (Nobel, 1970, p. 340), the entropy change in a typical photosynthetic reaction becomes

$$S_{phot} = 1.1 \times 10^{-9} \times (-34.55)\ \mathrm{J\ cm^{-2}\ s^{-1}\ K^{-1}}$$
$$= -0.38 \times 10^{-7}\ \mathrm{J\ cm^{-2}\ s^{-1}\ K^{-1}}$$

which is about 10^{-3} times the amount of the entropy flows thus far considered and hence can be neglected in the present consideration.

2.1.5 Net Entropy Flow and Entropy Production

The net entropy flow coming into the leaf from surroundings can be obtained by the summation of all the entropies thus far considered. The result is

$$\begin{aligned} S_{flow} &= (S_{dr} + S_{sc} + S_{rf}) \times \alpha + (S_{sky} + S_{grd}) \times \alpha_{IR} - S_{lf} - S_{evp} - S_{cnd} \\ &= -1.79 \times 10^{-4}\ \mathrm{J\ cm^{-2}\ s^{-1}\ K^{-1}} \end{aligned} \tag{2.28}$$

The negativeness of S_{flow} shows that the leaf absorbs "negative entropy" (Schrödinger, 1944) from surroundings. The negative value of the net entropy flow in Eq. (2.28) is thus the basis, from the entropy point of view, for the existence of organized structure and function in the plant leaf (Schrödinger, 1944). The net entropy flows from the leaf to surroundings due to IR radiation, the evaporation of water, and heat conduction are in the ratios 1.01:0.61:0.47 = 1.0:0.60:0.47.

If it is assumed that the leaf is in a steady state in entropy, as it should be from a short-term point of view, an entropy production S_{prod} should occur in the leaf so as to add to the entropy flow S_{flow} [Eq. (2.28)] to result in no change of total entropy:

$$S_{prod} = -S_{flow} = +1.79 \times 10^{-4}\ \mathrm{J\ cm^{-2}\ s^{-1}\ K^{-1}} \tag{2.29}$$

The formulation of the Second Law of Thermodynamics, extended to open systems, claims that the entropy production should always be positive [Eq. (1.2); Nicolis & Prigogine, 1977]. The positiveness of the entropy production in the leaf, shown in Eq. (2.29), is thus direct evidence that the Second Law of Thermodynamics is certainly valid in plant leaves. This is against the old arguments by Beier (1962) and Trincher (1967) that the Second Law cannot be applied to living organisms. Since entropy production can be regarded as a measure of the extent of strength of motions and reactions that occur in nature (Chapter 1; Aoki, 1983), it can be said that the motions and reactions, the extent of which is represented by the entropy production of the order of 10^{-4} J cm^{-2} s^{-1} K^{-1}, are necessary in order to maintain the organized activity and structure in the plant leaf.

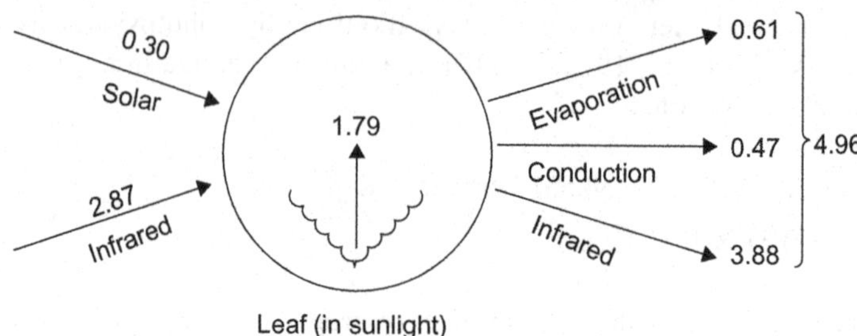

Figure 2.2 Flow and production diagram of entropy in a broad-plant leaf in units of 10^{-4} J cm^{-2} s^{-1} K^{-1} = J m^{-2} s^{-1} K^{-1}. The outgoing entropy 4.96 consists of the entropy due to IR radiation 3.88, that due to the evaporation of water 0.61 and that due to heat conduction 0.47. Other values are given in the text.

The entropy flow and entropy production thus far clarified are illustrated in Figure 2.2, a so-called flow-production diagram of entropy in the leaf. As shown, the entropy flux of solar shortwave radiation is smaller than that of terrestrial IR radiation and smaller than the entropy production in the leaf. The smallness of the entropy flow of solar radiation is necessary for the net entropy flow into the leaf to be negative and thus for the ordered structure and function of the leaf to be maintained.

2.1.6 Other Examples

Similar calculations can be done in other cases. The following are some examples.

Case I. A leaf at noon on a clear day at an altitude 2,000 m, where $E_i = 1.50$ cal cm^{-2} min^{-1} (second line of table 7.3 of Nobel, 1970): suppose that $d = 0.20$, $r = 0.20$, $\alpha = 0.60$, and $\alpha_{IR} = \varepsilon_{IR} = 0.96$. In this case, the entropy production in the leaf becomes

$$S_{prod} = 2.19 \times 10^{-4} \text{ J cm}^{-2} \text{ s}^{-1} \text{ K}^{-1} \tag{2.30}$$

Case II. A leaf at noon on a clear day in summer at an altitude 1,700 m (an example given by Gates, 1963):

$$S_{prod} = 2.78 \times 10^{-4} \text{ J cm}^{-2} \text{ s}^{-1} \text{ K}^{-1} \tag{2.31}$$

Case III. A silvery leaf ($\alpha = 0.50$) at $E_i = 1.50$ cal cm^{-2} min^{-1} (third line of table 7.3 of Nobel, 1970): suppose that $d = 0.20$, $r = 0.20$, and $\alpha_{IR} = \varepsilon_{IR} = 0.96$

$$S_{prod} = 1.89 \times 10^{-4} \text{ J cm}^{-2} \text{ s}^{-1} \text{ K}^{-1} \tag{2.32}$$

Case IV. A leaf on a cloudless night at sea level with frost formation (the bottom line of table 7.3 of Nobel, 1970, p. 360): in this case, heat flows due to frost formation and conduction are in a direction from the surroundings to the leaf. The heat flow due to frost formation is $0.05\ \mathrm{cal\ cm^{-2}\ min^{-1}}$, and the heat flow due to conduction is $0.10\ \mathrm{cal\ cm^{-2}\ min^{-1}}$ (Nobel, 1970, p. 360). Suppose that $\alpha_{IR} = \varepsilon_{IR} = 0.96$. In this case, the entropy production becomes

$$S_{prod} = 0.03 \times 10^{-4}\ \mathrm{J\ cm^{-2}\ s^{-1}\ K^{-1}} \tag{2.33}$$

Thus, by comparison with the entropy production in the leaf in mild and "standard" conditions [Eq. (2.29)], leaves with more of a solar radiation load (Cases I and II) produce much more entropy within them. When the leaf is silvery, even with high solar radiation load (Case III), more solar radiation is reflected, and the entropy production decreases nearly to the standard one. On a cloudless night with frost formation (Case IV), the entropy production becomes almost zero. This means that the IR radiation absorbed by leaves at night does not contribute to the entropy production in leaves.

2.1.7 Discussion

In the expressions in Eqs (2.10) and (2.15), the temperature of the Sun T_0 appears as a denominator; some readers may suggest that the temperature of the leaf should be used instead of T_0. For a detailed reasoning for these equations, readers should refer to the discussion fully developed in Aoki (1987b, 1988b). However, for the moment, a simple argument for this problem follows.

Let the Sun be regarded as a black-body sphere of a temperature T_0 and of a radius r. The radiation energy emitted per unit time from the unit area of the surface of the Sun is, according to the Stefan–Boltzmann Law (Planck, 1959, 1988), σT_0^4, and the corresponding radiation entropy emitted is $(4/3)\sigma T_0^3$, as already stated in connection with Eq. (2.18) (Planck, 1959, 1988). The total energy E emitted from the Sun per unit time is given by $E = 4\pi r^2 \sigma T_0^4$, and the total entropy S by $S = 4\pi r^2 (4/3)\sigma T_0^3 = (4/3)E/T_0$. If there are no energy losses in space, E equals the energy E_1 that passes in unit time through a larger sphere of a radius R ($>r$). Likewise, if the entropy of radiation does not change in space (i.e., if radiation is not absorbed, reflected, and scattered in space), then S equals the entropy S_1 that passes in unit time through the sphere. Let $E_1 = 4\pi R^2 e_1$ and $S_1 = 4\pi R^2 s_1$, where e_1 is the energy flux through the sphere and s_1 the corresponding entropy flux. From $E = E_1$ and $S = S_1$, e_1 and s_1 are expressed as: $e_1 = (r/R)^2 \sigma T_0^4$ and $s_1 = (r/R)^2 (4/3)\sigma T_0^3 = (4/3)e_1/T_0$. If R is the mean distance between the Sun and the Earth, then e_1 is the solar energy flux at the Earth, and s_1 is the corresponding entropy flux. Thus, in the expression for entropy flux at the Earth s_1, the temperature of the Sun T_0 appears (not the temperature of the Earth) as a denominator. With the factor (4/3), this is characteristic of *black-body radiation* entropy, which is not usually described in undergraduate textbooks on physics. The appearance of

$X(\varepsilon)$ in Eqs (2.10) and (2.15) is, respectively, due to the scattering of solar radiation by small particles in the air and the reflection of solar radiation by the ground. For a more detailed argument, refer to the appendix of Aoki (1987b), Aoki (1998), and the literatures cited therein.

The implications of these results, together with those in next two sections, are discussed in Section 2.4.

2.2 Conifer Branches (Aoki, 1989b)

The present section deals with entropy production in conifer branches (Aoki, 1989b). This is more complicated to treat than that in broad leaves because geometrical factors (such as the total surface area of a branch, the effective area of a branch absorbing direct solar radiation, the effective area of a branch absorbing radiation from extended sources, and the effective area for the emission of IR radiation) are more difficult to measure than in the case of broad-plant leaves. In this study, the geometrical factors given by Gates, Tibbals, and Kreith (1965) in their studies on energy budgets of conifers are adopted, and corresponding entropy fluxes and entropy productions in conifer branches are calculated using methods similar to those described in the previous section.

2.2.1 Solar Radiation Absorbed by a Branch of Ponderosa Pine

According to Gates et al. (1965), consider a branch of ponderosa pine exposed to global (total) solar radiation $E_i = 1.20\ \text{cal cm}^{-2}\ \text{min}^{-1} = 0.0837\ \text{J cm}^{-2}\ \text{s}^{-1}$, in which the energy of direct solar radiation is $E_{dr} = 1.05\ \text{cal cm}^{-2}\ \text{min}^{-1} = 0.0733\ \text{J cm}^{-2}\ \text{s}^{-1}$, and the energy of diffuse (scattered) solar radiation is $E_{sc} = 0.15\ \text{cal cm}^{-2}\ \text{min}^{-1} = 0.0105\ \text{J cm}^{-2}\ \text{s}^{-1}$. Also suppose that the reflectivity of solar radiation by the Earth's surface is $r = 0.10$ (Gates et al., 1965), that is, the energy of reflected solar radiation is $E_{rf} = E_i \times r = 0.12\ \text{cal cm}^{-2}\ \text{min}^{-1} = 0.0084\ \text{J cm}^{-2}\ \text{s}^{-1}$. The following values for the branch of ponderosa pine given by Gates et al. (1965) will also be adopted: the total surface area of the branch $A = 364\ \text{cm}^2$, the effective area of the branch absorbing direct solar radiation $A_{es} = 105\ \text{cm}^2$, the effective area absorbing radiation from extended sources and the effective area for the emission of IR radiation $A_{ee} = 310\ \text{cm}^2$, the absorptivity of direct solar radiation $\alpha_s = 0.70$, and the absorptivity of diffuse solar radiation $\alpha_s' = 0.80$.

The methods described in Section 2.1 (Aoki, 1987a, 1987b) are used without repeating the reasoning behind the techniques. First, consider direct solar radiation. The direct solar entropy flux incident on the branch of ponderosa pine is given by [see Eq. (2.2)]

$$\begin{aligned} S^0_{dr} &= 2.31 \times 10^{-4}\ \text{K}^{-1} \times E_{dr} \\ &= 0.170 \times 10^{-4}\ \text{J cm}^{-2}\ \text{s}^{-1}\ \text{K}^{-1} \end{aligned} \tag{2.34}$$

Hence, the direct solar entropy absorbed by the branch becomes

$$\begin{aligned} S_{dr} &= \alpha_s S^0_{dr} A_{es} = 0.70 \times 0.170 \times 10^{-4} \times 105 \\ &= 12.5 \times 10^{-4}\ \mathrm{J\ s^{-1}\ K^{-1}} \end{aligned} \tag{2.35}$$

Next, the specific intensity of diffuse solar radiation is given by Eq. (2.4)

$$K_1 = \frac{E_{sc}}{\pi} = 3.33 \times 10^{-3}\ \mathrm{J\ cm^{-2}\ s^{-1}} \tag{2.36}$$

On the other hand, the specific intensity of extraterrestrial solar radiation incident on the Earth is [Eq. (2.5)]

$$K_0 = 1.99 \times 10^3\ \mathrm{J\ cm^{-2}\ s^{-1}} \tag{2.37}$$

The "emissivity" of diffuse solar radiation is [Eq. (2.6)]

$$\varepsilon = \frac{K_1}{K_0} = 1.68 \times 10^{-6} \tag{2.38}$$

and the function $X(\varepsilon)$ becomes

$$X(\varepsilon = 1.68 \times 10^{-6}) = 4.65 \tag{2.39}$$

where [see Eqs (2.8) and (2.9)]

$$X(\varepsilon) = \frac{45}{4\pi^4}\frac{1}{\varepsilon}\int_0^{\infty} y^2[(x+1)\ln(x+1) - x\ln x]\mathrm{d}y \tag{2.40}$$

$$x = \frac{\varepsilon}{e^y - 1} \tag{2.41}$$

The entropy flux of diffuse solar radiation incident on the branch is [Eq. (2.10)]

$$S^0_{sc} = \frac{4}{3}\frac{E_{sc}}{T_0}X(\varepsilon) = 0.113 \times 10^{-4}\ \mathrm{J\ cm^{-2}\ s^{-1}\ K^{-1}} \tag{2.42}$$

where $T_0 = 5{,}760$ K is the temperature of the Sun. Since the effective area of the branch absorbing diffuse solar radiation is about $(A_{ee}/2)$ (Gates et al., 1965), the diffuse solar entropy absorbed by the branch becomes

$$\begin{aligned} S_{sc} &= \alpha'_s S^0_{sc}\,(A_{ee}/2) = 0.80 \times 0.113 \times 10^{-4} \times (310/2) \\ &= 14.0 \times 10^{-4}\ \mathrm{J\ s^{-1}\ K^{-1}} \end{aligned} \tag{2.43}$$

Similarly, the specific intensity of reflected solar radiation is [Eq. (2.12)]

$$K_1 = \frac{E_{rf}}{\pi} = 2.67 \times 10^{-3}\ \mathrm{J\ cm^{-2}\ s^{-1}} \tag{2.44}$$

Hence,

$$\varepsilon = \frac{K_1}{K_0} = 1.34 \times 10^{-6} \tag{2.45}$$

where K_0 is given by Eq. (2.5). The function $X(\varepsilon)$ becomes

$$X(\varepsilon = 1.34 \times 10^{-6}) = 4.70 \tag{2.46}$$

where $X(\varepsilon)$ is given by Eqs (2.40) and (2.41). The entropy flux of reflected solar radiation incident on the branch is [Eq. (2.15)]

$$S^0_{rf} = \frac{4}{3}\frac{E_{rf}}{T_0}X(\varepsilon) = 0.091 \times 10^{-4}\ \mathrm{J\ cm^{-2}\ s^{-1}\ K^{-1}} \tag{2.47}$$

where $T_0 = 5{,}760$ K is the temperature of the Sun. The reflected solar entropy absorbed by the branch becomes

$$\begin{aligned} S_{rf} &= \alpha'_s S^0_{rf}(A_{ee}/2) = 0.80 \times 0.091 \times 10^{-4} \times (310/2) \\ &= 11.3 \times 10^{-4}\ \mathrm{J\ s^{-1}\ K^{-1}} \end{aligned} \tag{2.48}$$

2.2.2 IR Radiation Entropy Absorbed by the Branch

Suppose that the IR energy flux from the sky to the branch is $E_{sky} = 0.48\ \mathrm{cal\ cm^{-2}\ min^{-1}} = 0.0335\ \mathrm{J\ cm^{-2}\ s^{-1}}$ (Gates et al., 1965). Let us regard IR radiation from the sky as black-body radiation. From the Stefan–Boltzmann Law, the effective temperature of the sky is given by

$$T_{sky} = \left[\frac{E_{sky}}{\sigma}\right]^{1/4} = 277\ \mathrm{K} \tag{2.49}$$

where σ is the Stefan–Boltzmann constant. The IR entropy flux from the sky is [Eq. (2.18)]

$$S^0_{sky} = \frac{4}{3}\sigma(T_{sky})^3 = \frac{4}{3}\frac{E_{sky}}{T_{sky}} = 1.61 \times 10^{-4}\ \mathrm{J\ cm^{-2}\ s^{-1}\ K^{-1}} \tag{2.50}$$

Suppose that the IR energy flux from the ground to the branch is $E_{grd} = 0.62$ cal cm^{-2} min^{-1} = 0.0433 J cm^{-2} s^{-1} (Gates et al., 1965). The IR radiation from the ground is regarded as black-body radiation. The effective temperature of the ground is

$$T_{grd} = \left[\frac{E_{grd}}{\sigma}\right]^{1/4} = 296\ \mathrm{K} \tag{2.51}$$

Hence, the IR entropy flux from the ground is

$$S^0_{grd} = \frac{4}{3}\sigma(T_{grd})^3 = \frac{4\,E_{grd}}{3\,T_{grd}} = 1.95 \times 10^{-4}\ \mathrm{J\ cm^{-2}\ s^{-1}\ K^{-1}} \tag{2.52}$$

Since the absorptivity of the branch toward IR radiation is $\alpha_t = 0.97$ (Gates et al., 1965), the total IR entropy absorbed by the branch is

$$\begin{aligned} S_{sky} + S_{grd} &= \alpha_t(S^0_{sky} + S^0_{grd})(A_{ee}/2) \\ &= 0.97 \times (1.61 + 1.95) \times 10^{-4} \times (310/2) \\ &= 535.6 \times 10^{-4}\ \mathrm{J\ s^{-1}\ K^{-1}} \end{aligned} \tag{2.53}$$

2.2.3 Temperature of the Branch and Entropy Associated with Transpiration

Let us suppose that the transpiration rate is 1.7×10^{-4} g cm^{-2} min^{-1} (Gates et al., 1965). Since the latent heat is 580 cal g^{-1}, the energy flux associated with transpiration is

$$\begin{aligned} E^0_{evp} &= 1.7 \times 10^{-4} \times 580 = 0.0986\ \mathrm{cal\ cm^{-2}\ min^{-1}} \\ &= 0.00688\ \mathrm{J\ cm^{-2}\ s^{-1}} \end{aligned} \tag{2.54}$$

Using the total surface area of the branch $A = 364$ cm^2, the transpiration energy from the branch is

$$E_{evp} = E^0_{evp} \times A = 35.9\ \mathrm{cal\ min^{-1}} = 2.50\ \mathrm{J\ s^{-1}} \tag{2.55}$$

The temperature difference ΔT between the temperature of the branch T_{lf} and that of the air T_{air} ($T_{air} = 25°$C according to Gates et al., 1965) can be obtained by solving the following energy balance equation:

$$274.3 = 189.0 + 35.9 + 6.52(\Delta T) + 1.31(\Delta T)^{1.3} \tag{2.56}$$

instead of the analogous equation

$$274.3 = 189.0 + 6.52(\Delta T) + 1.31(\Delta T)^{1.3} \tag{2.57}$$

given by Gates et al. (1965), in which no transpiration is considered. [The term 35.9 in Eq. (2.56) is the transpiration energy given by Eq. (2.55).] From Eq. (2.56), $\Delta T \simeq 5.7°\mathrm{C}$ and

$$T_{\mathrm{lf}} = T_{\mathrm{air}} + \Delta T = 30.7°\mathrm{C} = 303.8\ \mathrm{K} \tag{2.58}$$

The entropy flow associated with transpiration becomes

$$S_{\mathrm{evp}} = \frac{E_{\mathrm{evp}}}{T_{\mathrm{lf}}} = 82.4 \times 10^{-4}\ \mathrm{J\ s^{-1}\ K^{-1}} \tag{2.59}$$

2.2.4 Entropy Associated with Free Convection

The energy associated with free convection is obtained from the equation

$$E_{\mathrm{cnv}} = 4.00(\Delta T) + 1.31(\Delta T)^{1.3}\ \mathrm{cal\ min^{-1}} \tag{2.60}$$

obtained by Gates et al. (1965). By use of $\Delta T \simeq 5.7°\mathrm{C}$, Eq. (2.60) gives

$$E_{\mathrm{cnv}} = 35.4\ \mathrm{cal\ min^{-1}} = 2.47\ \mathrm{J\ s^{-1}} \tag{2.61}$$

The entropy flow associated with free convection becomes

$$S_{\mathrm{cnv}} = \frac{E_{\mathrm{cnv}}}{T_{\mathrm{lf}}} = 81.3 \times 10^{-4}\ \mathrm{J\ s^{-1}\ K^{-1}} \tag{2.62}$$

2.2.5 IR Radiation Entropy Emitted by the Branch

Since the emissivity of the branch for IR radiation is $\varepsilon_{\mathrm{t}} = 0.97$ (Gates et al., 1965), the function $X(\varepsilon_{\mathrm{t}})$ becomes

$$X(\varepsilon_{\mathrm{t}} = 0.97) = 1.008 \tag{2.63}$$

where $X(\varepsilon_{\mathrm{t}})$ is given by Eqs (2.40) and (2.41). The entropy flux emitted by the plant surface is [Eq. (2.25)]

$$\begin{aligned} S^0_{\mathrm{lf}} &= \frac{4}{3}\varepsilon_{\mathrm{t}}\sigma(T_{\mathrm{lf}})^3 X(\varepsilon_{\mathrm{t}}) \\ &= 2.07 \times 10^{-4}\ \mathrm{J\ cm^{-2}\ s^{-1}\ K^{-1}} \end{aligned} \tag{2.64}$$

Hence, the IR entropy emitted by the branch is obtained as

$$S_{\rm lf} = S_{\rm lf}^{0} \times A_{\rm ee} = 642.5 \times 10^{-4}\ {\rm J\ s^{-1}\ K^{-1}} \tag{2.65}$$

2.2.6 Net Entropy Flow and Entropy Production

The net entropy flow from the surroundings into the branch is obtained as

$$\begin{aligned} S_{\rm flow} &= S_{\rm dr} + S_{\rm sc} + S_{\rm rf} + S_{\rm sky} + S_{\rm grd} - S_{\rm lf} - S_{\rm evp} - S_{\rm cnv} \\ &= -232.9 \times 10^{-4}\ {\rm J\ s^{-1}\ K^{-1}} \end{aligned} \tag{2.66}$$

Thus, the branch of ponderosa pine absorbs "negative entropy" (Schrödinger, 1944) from its surroundings, as is the case for broad-plant leaves (Section 2.1). From the entropy point of view, this is the basis for the organized functions and structures in the branch to be maintained (Schrödinger, 1944). The net entropy flows from the branch to surroundings due to IR radiation, transpiration, and convection are in the ratios 1.0:0.8:0.8.

If the branch is at a steady state in entropy, an entropy production $S_{\rm prod}$ should occur in the branch so as to add to the entropy flow $S_{\rm flow}$, to result in no change of total entropy:

$$S_{\rm prod} = -S_{\rm flow} = +232.9 \times 10^{-4}\ {\rm J\ s^{-1}\ K^{-1}} \tag{2.67}$$

The positiveness of $S_{\rm prod}$ shows that the Second Law of Thermodynamics holds in the branch, as in the case of broad-plant leaves (Section 2.1). Since the total branch surface is $A = 364\ {\rm cm}^2$, the entropy production per unit area of the plant surface becomes

$$\frac{S_{\rm prod}}{A} = 0.64 \times 10^{-4}\ {\rm J\ cm^{-2}\ s^{-1}\ K^{-1}} \tag{2.68}$$

As already shown, the value of $1.7 \times 10^{-4}\ {\rm g\ cm^{-2}\ min^{-1}}$ for the transpiration rate of ponderosa pine has been adopted. However, changes of the value of transpiration rate have little or almost no effect on the result for entropy production, as shown in the following discussion. If the transpiration rate increases, the value of $E_{\rm evp}$ increases, and hence ΔT decreases by Eq. (2.56) and $T_{\rm lf}$ decreases by Eq. (2.58). Then, the value of $S_{\rm evp}$ increases [Eq. (2.59)]. On the other hand, the decrease of ΔT leads to the decrease of $E_{\rm cnv}$ [Eq. (2.60)] and hence of $S_{\rm cnv}$. Also, the decrease of $T_{\rm lf}$ leads to the decrease of $S_{\rm lf}$ [Eq. (2.64)]. By numerical calculations, it is shown that most of the increment of $S_{\rm evp}$ is canceled, mainly by the decrement of $S_{\rm cnv}$ and a little by the decrement of $S_{\rm lf}$. Then, the change of transpiration rate induces only a small change (less than the last significant figure) for the sum $(S_{\rm evp} + S_{\rm cnv} + S_{\rm lf})$ and hence for the entropy production. For example, the increase of transpiration rate by 50% results in the decrease of the entropy production by only 0.2%. Thus, the change of transpiration rate has almost no effect on entropy production. This is also the case for broad leaves, although it was not explicitly stated in the previous section.

2.2.7 Other Cases

Similar calculations can be carried out for other conifer branches. The transpiration rate is assumed to be 1.7×10^{-4} g cm^{-2} min^{-1} for all cases. The results are not largely influenced by transpiration rate, as stated.

Case I. A branch of blue spruce: $A = 124$ cm^2, $A_{es} = 42$ cm^2, $A_{ee} = 109$ cm^2, $\alpha_s = 0.70$, $\alpha'_s = 0.80$, $\alpha_t = \varepsilon_t = 0.97$ as reported by Gates et al. (1965). The entropy production per unit area of the plant surface becomes

$$\frac{S_{prod}}{A} = 0.73 \times 10^{-4}\ \mathrm{J\,cm^{-2}\,s^{-1}\,K^{-1}} \tag{2.69}$$

Case II. A branch of white fir: $A = 109$ cm^2, $A_{es} = 38.3$ cm^2, $A_{ee} = 108$ cm^2, $\alpha_s = 0.70$, $\alpha'_s = 0.80$, $\alpha_t = \varepsilon_t = 0.97$ (Gates et al., 1965). The entropy production per unit area of the plant surface becomes

$$\frac{S_{prod}}{A} = 0.77 \times 10^{-4}\ \mathrm{J\,cm^{-2}\,s^{-1}\,K^{-1}} \tag{2.70}$$

Case III. A single pine needle in a horizontal position: The radius of the needle is $r = 0.1$ cm, and the length is l; $A = 2\pi rl$, $A_{es} = 2rl$, $A_{ee} = 2\pi rl$ (Gates et al., 1965). In this case,

$$\frac{S_{prod}}{A} = 0.70 \times 10^{-4}\ \mathrm{J\,cm^{-2}\,s^{-1}\,K^{-1}} \tag{2.71}$$

These values of entropy production are plotted against absorbed solar energy in Section 2.4.

2.3 Entropy Production at Night

Energy flows in soybean leaves were observed and estimated on freezing nights by Schwintzer (1971) and in bur oak leaves on warm nights by Gates (1964). By the use of similar methods as in Sections 2.1 and 2.2, entropy productions in these leaves at nights are calculated (Aoki, 1987c). The results show that entropy production in these leaves at nights is almost zero, which is consistent with a result in Section 2.1. Night respiration is too small to be observed by the present methods.

2.4 Theorem of Oscillating Entropy Production

The solar radiation energy absorbed by the leaf surface is given by $E_{solar} = (E_{dr} + E_{sc} + E_{rf}) \times \alpha$ for broad-plant leaves in Section 2.1 and is also given in Gates et al. (1965) for conifer branches. The results of entropy production for broad-plant leaves (Section 2.1), conifer branches (Section 2.2), and leaves at night (Section 2.3), plotted against the absorbed solar energy, are shown in Figure 2.3. Thus, entropy production and absorbed solar radiation energy are in a linear relationship:

$$S_{prod} \simeq (30.6 E_{solar}) \times 10^{-4} \tag{2.72}$$

There is no difference in the sensitivity of entropy production to absorbed solar energy between broad-plant leaves and conifer branches. The nature of the physical–chemical interaction of solar radiation to the internal structure of both plant leaves are same and may be universal for all plants.

From the linearity of the entropy production with the absorbed solar energy, which oscillates from day to day, the entropy production should also oscillate with a period of 1 day (the oscillation theorem of entropy production in plants).

The amplitude of oscillation becomes large over the growing period of leaves, reaches its maximum at the adult steady state, and becomes small as the senescent

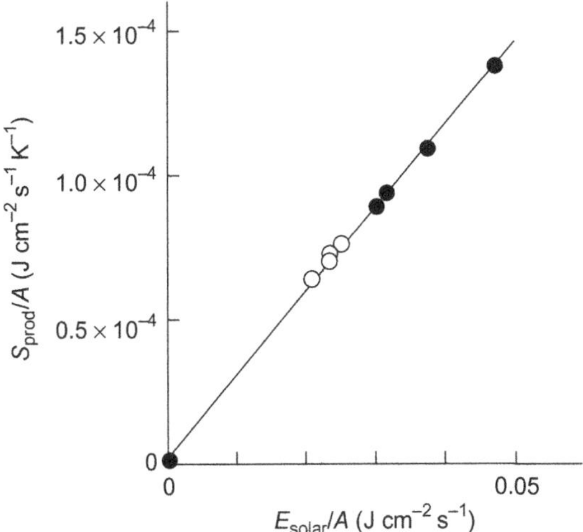

Figure 2.3 The entropy production per unit area of the plant surface (S_{prod}/A) plotted against the solar energy absorbed by a unit area of the surface (E_{solar}/A). ●: Broad leaves (Sections 2.1 and 2.3); ○: conifer branches (Section 2.2). *Note*: S_{prod} and E_{solar}, given in Section 2.1, are the values per both sides of the surface of broad leaves; hence they are divided by the factor 2 here.

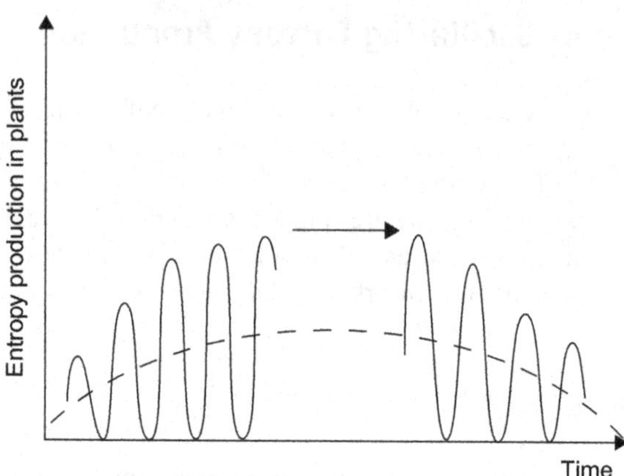

Figure 2.4 Schematic profile of entropy production in plants versus time, showing daily oscillation. The dashed line is an average of the amplitude of oscillation, which tends to increase in the growing state, to stay steady state in the adult stage, and to decrease in the senescent stage.

period progresses (even the leaves of conifers undergo senescence and death). Over the life history of leaves, the average of the amplitude shows a tendency of initial increase, intermediate steadiness, and later a decrease (Figure 2.4). (This multiphase character is the key concept of the development of living systems from organisms to ecosystems, as shown in later chapters.) These results apply to leaves and branches and may also be applicable to forests over a long time span.

3 Animals

The extensive comparison of the entropy production of many species of animals may lead to some area of research, such as entropy allometry or entropy scaling law, in addition of that of energy (Kleiber, 1975) (entropy production per surface area/body weight versus body weight). However, the data are not enough at present (a preliminary report was presented by Aoki, 1985). The entropy productions of only three animals are shown in this chapter: deer (Aoki, 1987d), lizard (Aoki, 1988a), and pig (Aoki, 1992). A white-tailed deer on a winter night is the first in which entropy production has been quantitatively determined in animals because calculation is simple: solar radiation entropy is not considered. This is not the case in the plant leaves in Chapter 2 and in a lizard in sunlight in a later section. Nonphysics readers may skip the mathematical parts of Section 3.2. Also, the time course of entropy production in the growth of a pig is presented here; a similar approach may be seen in later chapters.

3.1 White-Tailed Deer

Moen (1966, 1968a, 1968b, 1973) intensively studied the energetics of white-tailed deer (*Odocoileus virginianus*) in an open field environment under clear nocturnal skies in winter. A natural extension of this line of research (energetics) will be an investigation of white-tailed deer from an entropy point of view (entropetics), as is evident from the formulation of thermodynamics: the First Law of Thermodynamics is concerned with energy, and the Second Law with entropy. In this section, entropy flow and entropy production in a white-tailed deer (50 kg) on a maintenance diet and a full-feed diet in a standing posture in an open field under clear nocturnal skies with an air temperature of $-20°C$ are calculated based on the energetics given by Moen (1968a, 1968b, 1973) and Aoki (1987d).

The basal metabolic rate (BMR) used here is 1.292 kcal/kg/h for a fasting and lying white-tailed deer (Moen, 1968b). The ratio of heat production on a maintenance diet to that on a fasting diet is 1.47 : 1, and the ratio of heat production in a standing to a lying posture is 1.1 : 1 (Moen, 1968b). Hence, the metabolic heat production of a 50 kg deer on a maintenance diet in a standing posture is

$$\begin{aligned} E_{\text{mtb}} &= 1.292 \times 50 \times 1.47 \times 1.1 \text{ kcal h}^{-1} \\ &= 121.5 \text{ J s}^{-1} \end{aligned} \tag{3.1}$$

Entropy Principle for the Development of Complex Biotic Systems. DOI: 10.1016/B978-0-12-391493-4.00003-2

Since the body temperature of a deer is $T_b = 39°C = 312.2$ K (Moen, 1968b), the entropy production due to metabolic heat production becomes

$$S_{mtb} = \frac{E_{mtb}}{T_b} = 0.389 \text{ J s}^{-1} \text{ K}^{-1} \tag{3.2}$$

Infrared (IR) radiation from the sky and from the ground is incident on and absorbed by a white-tailed deer. Downward and upward IR radiation energy fluxes in open fields under clear night skies are, respectively (Moen, 1973, p. 81, or Moen & Evans, 1971)

$$E\downarrow = \sigma[(-8.920 + 1.10T_a) + 273.2]^4$$

and

$$E\uparrow = \sigma[(-0.049 + 1.03T_a) + 273.2]^4 \tag{3.3}$$

where σ is the Stefan–Boltzmann constant, and T_a is the air temperature (°C). That is, the effective temperature of the sky is $T_{sky} = (-8.920 + 1.10T_a) + 273.2$, and that of the ground is $T_{grd} = (-0.049 + 1.03T_a) + 273.2$. The radiation profile of a deer is assumed to be 0.85 (Moen, 1968b). The total surface area S_t of a 50 kg female deer is 1.703 m^2, obtained from the regression equation $S_t = 0.142$ (weight in kg)$^{0.635}$ (Moen, 1973, p. 437). The IR absorptivity (equals the IR emissivity) is assumed to be 1 (Moen, 1968a). When $T_a = -20°C$, the IR radiation energy absorbed by a white-tailed deer becomes

$$\begin{aligned} E_{rad,in} &= 0.85 \times S_t \times \frac{1}{2}(E\downarrow + E\uparrow) \\ &= 308.4 \text{ J s}^{-1} \end{aligned} \tag{3.4}$$

[see also Eq. (5) of Moen (1968b)].

Entropies associated with these IR radiation can be calculated in the following way. The effective temperature of the sky is $T_{sky} = (-8.920 + 1.10T_a) + 273.2$, as already stated. When $T_a = -20°C$, the downward entropy flux is [for radiation entropy, refer to Planck (1959, 1988) or to Aoki (1982a, 1982b, 1983, 1987b, 1998)]

$$S\downarrow = \frac{4}{3}\sigma(T_{sky})^3 = 1.075 \text{ J m}^{-2} \text{ s}^{-1} \text{ K}^{-1} \tag{3.5}$$

Similarly, the effective temperature of the ground is $T_{grd} = (-0.049 + 1.03T_a) +$ 273.2, and the upward entropy flux is, for $T_a = -20°C$,

$$S\uparrow = \frac{4}{3}\sigma(T_{grd})^3 = 1.218 \text{ J m}^{-2} \text{ s}^{-1} \text{ K}^{-1} \tag{3.6}$$

The IR radiation entropy absorbed by a white-tailed deer becomes

$$\begin{aligned} S_{\mathrm{rad,in}} &= 0.85 \times S_{\mathrm{t}} \times \frac{1}{2}(S\downarrow + S\uparrow) \\ &= 1.660\ \mathrm{J\ s^{-1}\ K^{-1}} \end{aligned} \tag{3.7}$$

[cf. Eq. (3.4)].

Two modes of energy gain by a deer have just been described. Next, let us consider energy loss from a white-tailed deer. The average energy gain must balance with the energy loss for the deer to maintain a steady state in energy. If energy gain is not enough to balance a net loss of energy, body temperature must drop, and a deer will eventually die if the imbalance continues. If energy gain is greater than energy loss, the body temperature must increase, and, again, a deer will die if the increase continues. From figure 2 of Moen (1968b), the energy gain balances the energy loss when wind velocity at deer height is 4 mph for a 50 kg deer on a maintenance diet in a standing posture in an open field under clear nocturnal skies with an air temperature of −20°C. At that wind velocity, the percentages of energy lost by four modes of energy transfer from a 50 kg deer at −20°C are 33.3% of $E_{\mathrm{mtb}} = 121.5\ \mathrm{J\ s^{-1}}$ for net radiation loss, 44.0% for convective heat loss, 17.0% for evaporative heat loss, and 5.7% for conductive heat loss to ingested food, as can be seen in figure 3 of Moen (1968b). That is, net radiation loss is $E_{\mathrm{rad,net}} = 40.49\ \mathrm{J\ s^{-1}}$, convective heat loss is $E_{\mathrm{cnv}} = 53.44\ \mathrm{J\ s^{-1}}$, evaporative heat loss is $E_{\mathrm{evp}} = 20.65\ \mathrm{J\ s^{-1}}$, and conductive heat loss to ingested food is $E_{\mathrm{cnd}} = 6.88\ \mathrm{J\ s^{-1}}$.

IR radiation energy emitted by a deer is $E_{\mathrm{rad,out}} = E_{\mathrm{rad,in}} + E_{\mathrm{rad,net}} = 348.8\ \mathrm{J\ s^{-1}}$ [see Eq. (3.4)]. Since this is expressed as $E_{\mathrm{rad,out}} = 0.85 \times S_{\mathrm{t}} \times \sigma(T_{\mathrm{s}})^4$, where T_{s} is the surface temperature of a white-tailed deer, the quantity T_{s} is given by

$$T_{\mathrm{s}} = \left(\frac{E_{\mathrm{rad,out}}}{0.85 \times S_{\mathrm{t}} \times \sigma}\right)^{1/4} = 255.3\ \mathrm{K} \tag{3.8}$$

Then, the IR entropy emitted by a deer is obtained as

$$\begin{aligned} S_{\mathrm{rad,out}} &= 0.85 \times S_{\mathrm{t}} \times \frac{4}{3}\sigma(T_{\mathrm{s}})^3 \\ &= 1.822\ \mathrm{J\ s^{-1}\ K^{-1}} \end{aligned} \tag{3.9}$$

The convective entropy loss associated with the convective heat loss becomes

$$S_{\mathrm{cnv}} = E_{\mathrm{cnv}} T_{\mathrm{s}}^{-1} = 0.209\ \mathrm{J\ s^{-1}\ K^{-1}} \tag{3.10}$$

Similarly, the evaporative entropy loss associated with the evaporative heat loss is given by

$$S_{\mathrm{evp}} = E_{\mathrm{evp}} T_{\mathrm{b}}^{-1} = 0.066\ \mathrm{J\ s^{-1}\ K^{-1}} \tag{3.11}$$

and the conductive entropy loss to ingested food is

$$S_{cnd} = E_{cnd} T_b^{-1} = 0.022 \text{ J s}^{-1} \text{ K}^{-1} \tag{3.12}$$

Net entropy flow into a white-tailed deer from its environment becomes

$$\begin{aligned} S_{flow} &= S_{rad,in} - [S_{rad,out} + S_{cnv} + S_{evp} + S_{cnd}] \\ &= -0.46 \text{ J s}^{-1} \text{ K}^{-1} \end{aligned} \tag{3.13}$$

That is, a white-tailed deer absorbs "negative entropy" (Schrödinger, 1944) from its surroundings. The physical basis to be maintained for organized structure in the body of a white-tailed deer (Schrödinger, 1944) is the same as in plant leaves (Chapter 2).

If it is assumed that a white-tailed deer is in a steady state in entropy as in energy, an entropy production S_{prod} should occur in a white-tailed deer so as to result in no change of total entropy:

$$S_{prod} = +0.46 \text{ J s}^{-1} \text{ K}^{-1} \tag{3.14}$$

Positiveness of the entropy production S_{prod} shows that the Second Law of Thermodynamics certainly holds in a 50 kg deer on a maintenance diet (Chapter 1). Of the total entropy production S_{prod}, the entropy production due to metabolic heat production was already given as $S_{mtb} = 0.39 \text{ J s}^{-1} \text{ K}^{-1}$. The remainder of the entropy production $0.07 \text{ J s}^{-1} \text{ K}^{-1}$ is due to irreversible motions and reactions of substances in a white-tailed deer not associated with metabolic heat production. The entropy production per effective radiating surface area is

$$\begin{aligned} \frac{S_{prod}}{0.85\, S_t} &= 0.32 \text{ J m}^{-2} \text{ s}^{-1} \text{ K}^{-1} \\ &= 0.32 \times 10^{-4} \text{ J cm}^{-2} \text{ s}^{-1} \text{ K}^{-1} \end{aligned} \tag{3.15}$$

and that for per body weight is

$$\begin{aligned} \frac{S_{prod}}{50 \text{ kg}} &= 0.0092 \text{ J kg}^{-1} \text{ s}^{-1} \text{ K}^{-1} \\ &= 0.092 \times 10^{-4} \text{ J g}^{-1} \text{ s}^{-1} \text{ K}^{-1} \end{aligned} \tag{3.16}$$

Figure 3.1 shows a flow-production diagram of entropy in a white-tailed deer on a maintenance diet and summarizes the calculation results thus far.

Similar calculations are also carried out for a 50 kg white-tailed deer on a full-feed diet in a standing posture in an open field under clear skies at night with the air temperature at −20°C. The entropy production in a white-tailed deer in this case becomes $S_{prod} = +0.59 \text{ J s}^{-1} \text{ K}^{-1}$. The entropy production per effective

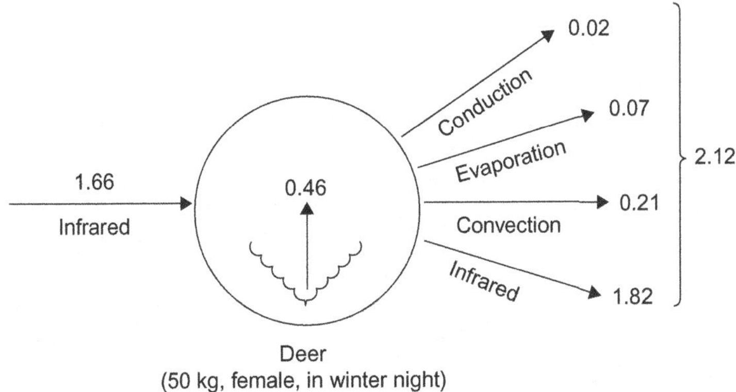

Figure 3.1 Entropy flow and entropy production for a white-tailed deer during a winter night. Units are $\mathrm{J\,s^{-1}\,K^{-1}}$.

radiating surface area is $0.41 \times 10^{-4}\ \mathrm{J\,cm^{-2}\,s^{-1}\,K^{-1}}$ and per body weight, it is $0.12 \times 10^{-4}\ \mathrm{J\,g^{-1}\,s^{-1}\,K^{-1}}$. Thus, a white-tailed deer on a full-feed diet produces more entropy than one on a maintenance diet; the difference between the entropy productions on a full-feed diet and a maintenance diet is $0.13\ \mathrm{J\,s^{-1}\,K^{-1}}$.

The levels of a maintenance diet and of a full-feed diet are (4 lb of food/100 lb of body weight) and (5 lb of food/100 lb of body weight) per day, respectively (Moen, 1968a). Hence, for a white-tailed deer of 50 kg, the dietary levels are 2 kg of food per day for a maintenance diet and 2.5 kg of food per day for a full-feed diet; the difference is 0.5 kg of food per day, or $0.0058\ \mathrm{g\,s^{-1}}$ food. Therefore, the difference in the entropy production related to the difference in the food uptake is $0.13\ \mathrm{J\,s^{-1}\,K^{-1}}/0.0058\ \mathrm{g\,s^{-1}} = 22\ \mathrm{J\,g^{-1}\,K^{-1}}$. That is, the additional uptake of 1 g of food over a maintenance diet produces $22\ \mathrm{J\,K^{-1}}$ of more entropy within the body of a white-tailed deer.

The entropy production in plant leaves is almost zero at night (Chapter 2). However, the entropy production in a white-tailed deer is not zero at night, as shown in this section. This is a natural consequence because physical and biochemical reactions in animals are active even at night. [Note that entropy production is a measure of the degree of motions and reactions of substances occurring in nature (Chapter 1).]

If the energetic data of infant, adult, and senescent white-tailed deer are available, the corresponding entropy productions would be obtained, and the time dependence of entropy production in the life span of white-tailed deer would be clear. However, data are not available at present.

This is the first time that entropy production has been numerically determined in animals. The calculation is simple because it contains no discussion of solar radiation entropy. If it should be included, discussions become more complex, as in plant leaves (Chapter 2) and in a lizard under sunlight (the next section).

3.2 Lizard

To clarify the entropy characteristics of living systems, it is important to make comparative studies of entropy production in many species of organisms under various environments. In this section, entropy flows and an entropy production for a lizard under sunlight in summer are calculated (Aoki, 1988a) and then compared with the results on a white-tailed deer during a winter night (Section 3.1). The calculations are based on the energetics of a lizard given by Bartlett and Gates (1967), in which the energy budget is studied for a lizard (*Sceloporus occidentalis*) located vertically on the south side of a tree trunk at 13:00 on June 21 on Chew's Ridge in the Los Padres National Forest, California. The method of calculation is described in Sections 2.1 and 2.2 and is used here without further explanation to avoid duplication.

3.2.1 Entropy Flow due to Direct Solar Radiation

First, let us consider the direct solar entropy incident on a lizard. The direct solar energy incident on a plane perpendicular to the direct solar radiation at 13:00 on June 21 on Chew's Ridge is (Bartlett & Gates, 1967)

$$\begin{aligned} E_{dr} &= 1.58 \text{ cal cm}^{-2} \text{ min}^{-1} \\ &= 1102 \text{ J m}^{-2} \text{ s}^{-1} \end{aligned} \tag{3.17}$$

Therefore, the corresponding solar entropy flux directly incident on the lizard is given by [see Eq. (2.2) or Eq. (A10) of Aoki, 1987b]

$$\begin{aligned} S_{dr} &= 2.31 \times 10^{-4} \text{ K}^{-1} \times E_{dr} \\ &= 0.255 \text{ J m}^{-2} \text{ s}^{-1} \text{ K}^{-1} \end{aligned} \tag{3.18}$$

According to Bartlett and Gates (1967), the absorptivity of the lizard to direct solar radiation is $\alpha_s = 0.885$, and the lizard surface area available for absorbing direct solar radiation is $A_s = 6.2 \times 10^{-4}$ m^2. Hence, the direct solar entropy absorbed by the lizard is

$$\begin{aligned} S^*_{dr} &= \alpha_s S_{dr} A_s \\ &= 0.885 \times 0.255 \times 6.2 \times 10^{-4} \text{ J s}^{-1} \text{ K}^{-1} \\ &= 1.4 \times 10^{-4} \text{ J s}^{-1} \text{ K}^{-1} \end{aligned} \tag{3.19}$$

3.2.2 Entropy Flow due to Diffuse Solar Radiation

Next, consider the entropy flow into the lizard associated with diffuse (scattered) solar radiation. The diffuse solar energy incident on the lizard that is located vertically on a tree trunk is (Bartlett & Gates, 1967)

$$E_{sc} = 0.04 \text{ cal cm}^{-2} \text{ min}^{-1} \\ = 27.9 \text{ J m}^{-2} \text{ s}^{-1} \quad (3.20)$$

In the following discussion, let us follow a line described in the appendix of Aoki (1987b). Suppose that a beam of diffuse solar radiation is incident on an element of area $d\sigma$ on a vertical plane on a big tree trunk or a fence through a solid angle $d\Omega = \sin\theta\, d\theta\, d\varphi$ in a direction forming an angle θ with the normal to the area $d\sigma$ (φ is an azimuthal angle of the incident radiation beam). Let the specific intensity (Planck, 1959, 1988) of diffuse solar radiation be denoted as K_1. The radiation energy incident on $d\sigma$ in time dt through $d\Omega$ is, by definition (Planck, 1959, 1988), $K_1\, dt\, d\sigma \cos\theta\, d\Omega$. Now, assume that diffuse solar radiation comes from all directions in the sky with equal intensity and that hence K_1 is independent of (θ, φ). Integrating $K_1\, dt\, d\sigma \cos\theta\, d\Omega$ by $d\Omega$, the total radiation energy incident on $d\sigma$ in time dt is obtained:

$$K_1\, dt\, d\sigma \int_0^{\pi} d\phi \int_0^{\pi/2} \cos\theta \sin\theta\, d\theta = \frac{1}{2}\pi K_1\, dt\, d\sigma \quad (3.21)$$

Thus, the energy of diffuse (scattered) solar radiation incident per unit time on a unit area of a vertical surface is expressed as

$$E_{sc} = \frac{1}{2}\pi K_1 \quad (3.22)$$

From Eqs (3.20) and (3.22),

$$K_1 = \frac{2}{\pi} E_{sc} = 17.8 \text{ J m}^{-2} \text{ s}^{-1} \quad (3.23)$$

However, the specific intensity of extraterrestrial solar radiation is [Eq. (2.5) or Eq. (B3) of Aoki, 1987b]

$$K_0 = 1.99 \times 10^7 \text{ J m}^{-2} \text{ s}^{-1} \quad (2.24)$$

Then, the "emissivity" ε of diffuse solar radiation is expressed as [Eq. (2.6) or Eq. (B4) of Aoki, 1987b]

$$\varepsilon = \frac{K_1}{K_0} = 0.894 \times 10^{-6} \quad (3.25)$$

and the function $X(\varepsilon)$ becomes

$$X(0.894 \times 10^{-6}) = 4.82 \quad (3.26)$$

where

$$
\begin{aligned}
X(\varepsilon) &= \frac{45}{4\pi^4}\frac{1}{\varepsilon}\int_0^{\infty} y^2[(1+x)\ln(1+x) - x\ln x]\mathrm{d}y \\
x &= \frac{\varepsilon}{\mathrm{e}^y - 1}
\end{aligned}
\tag{3.27}
$$

[Eqs (2.8) and (2.9) or Eqs (B6) and (B7) of Aoki, 1987b].

The specific intensity of diffuse entropy radiation is given by [Eq. (B8) of Aoki, 1987b]

$$
L_1 = \frac{4}{3}\frac{K_1}{T_0}X(\varepsilon) \tag{3.28}
$$

where $T_0 = 5{,}760$ K is the temperature of the sun. From a similar discussion as in Eq. (3.22), the entropy of diffuse solar radiation incident per unit time on a unit area of a vertical surface is expressed as

$$
S_{\mathrm{sc}} = \frac{1}{2}\pi L_1 \tag{3.29}
$$

From Eqs (3.22), (3.28), and (3.29),

$$
\begin{aligned}
S_{\mathrm{sc}} &= \frac{4}{3}\left(\frac{1}{2}\pi K_1\right)\frac{1}{T_0}X(\varepsilon) \\
&= \frac{4}{3}\frac{E_{\mathrm{sc}}}{T_0}X(\varepsilon)
\end{aligned}
\tag{3.30}
$$

By the use of $T_0 = 5{,}760$ K and Eqs (3.20) and (3.26),

$$
S_{\mathrm{sc}} = 0.0311\ \mathrm{J\,m^{-2}\,s^{-1}\,K^{-1}} \tag{3.31}
$$

According to Bartlett and Gates (1967), the absorptivity of the lizard to diffuse solar radiation is $\alpha_s = 0.90$, and the effective absorbing area (or emitting area) of the lizard not in contact with the tree is $A_e = 57.4 \times 10^{-4}\ \mathrm{m}^2$. Hence, the diffuse solar entropy absorbed by the lizard is

$$
\begin{aligned}
S^*_{\mathrm{sc}} &= \alpha_s S_{\mathrm{sc}} A_e \\
&= 0.90 \times 0.0311 \times 57.4 \times 10^{-4}\ \mathrm{J\,s^{-1}\,K^{-1}} \\
&= 1.6 \times 10^{-4}\ \mathrm{J\,s^{-1}\,K^{-1}}
\end{aligned}
\tag{3.32}
$$

3.2.3 Entropy Flow due to Reflected Solar Radiation

The reflectance of the ground surface to solar radiation is $r = 0.05$ (Bartlett & Gates, 1967). Hence, the solar energy reflected by the ground and incident on the lizard is

$$
\begin{aligned}
E_{\mathrm{rf}} &= r \times (E_{\mathrm{dr}} + E_{\mathrm{sc}}) \\
&= 0.05 \times (1.58 + 0.04)\ \mathrm{cal\ cm^{-2}\ min^{-1}} \\
&= 56.5\ \mathrm{J\ m^{-2}\ s^{-1}}
\end{aligned} \tag{3.33}
$$

Let us follow a discussion similar to that in Section 3.2.2. Suppose that reflected solar radiation is incident on an element of area $\mathrm{d}\sigma$ on a vertical plane on a big tree trunk or a fence through a solid angle $\mathrm{d}\Omega = \sin\theta\ \mathrm{d}\theta\ \mathrm{d}\varphi$ in a direction forming an angle θ with the normal to the area $\mathrm{d}\sigma$ (φ is an azimuthal angle of reflected radiation). Let the specific intensity (Planck, 1959, 1988) of reflected solar radiation be denoted as K_1. The reflected radiation energy incident on $\mathrm{d}\sigma$ in time $\mathrm{d}t$ through $\mathrm{d}\Omega$ is, by definition (Planck, 1959, 1988), $K_1\ \mathrm{d}t\ \mathrm{d}\sigma \cos\theta\ \mathrm{d}\Omega$. Let us assume that K_1 is independent of (θ, φ). Integrating $K_1\ \mathrm{d}t\ \mathrm{d}\sigma \cos\theta\ \mathrm{d}\Omega$ by $\mathrm{d}\Omega$, the total reflected radiation energy incident on $\mathrm{d}\sigma$ in time $\mathrm{d}t$ is obtained as

$$
K_1\ \mathrm{d}t\ \mathrm{d}\sigma \int_0^{\pi} \mathrm{d}\varphi \int_0^{\pi/2} \cos\theta \sin\theta\ \mathrm{d}\theta = \frac{1}{2}\pi K_1\ \mathrm{d}t\ \mathrm{d}\sigma \tag{3.34}
$$

Thus, the energy of reflected solar radiation incident per unit time on a unit area of a vertical surface is expressed as

$$
E_{\mathrm{rf}} = \frac{1}{2}\pi K_1 \tag{3.35}
$$

From Eqs (3.33) and (3.35),

$$
\begin{aligned}
K_1 &= \frac{2}{\pi} E_{\mathrm{rf}} \\
&= 36.0\ \mathrm{J\ m^{-2}\ s^{-1}}
\end{aligned} \tag{3.36}
$$

By the use of Eqs (3.24) and (3.36), the "emissivity" of reflected solar radiation is obtained as

$$
\begin{aligned}
\varepsilon &= \frac{K_1}{K_0} \\
&= 1.81 \times 10^{-6}
\end{aligned} \tag{3.37}
$$

and the function $X(\varepsilon)$ becomes

$$
X(1.81 \times 10^{-6}) = 4.63 \tag{3.38}
$$

where $X(\varepsilon)$ is given by Eq. (3.27).

From a discussion similar to that in Section 3.2.2 (see also Aoki, 1987b), the entropy of reflected solar radiation incident per unit time on a unit area of a vertical surface is expressed as [Eq. (C7) of Aoki, 1987b]

$$S_{rf} = \frac{4}{3}\frac{E_{rf}}{T_0}X(\varepsilon) \tag{3.39}$$

where $T_0 = 5{,}760$ K is the temperature of the sun. From $T_0 = 5{,}760$ K and from Eqs (3.33) and (3.38),

$$S_{rf} = 0.0606 \text{ J m}^{-2} \text{ s}^{-1} \text{ K}^{-1} \tag{3.40}$$

Hence, the reflected solar entropy absorbed by the lizard becomes

$$\begin{aligned} S^*_{rf} &= \alpha_s S_{rf} A_e \\ &= 0.90 \times 0.0606 \times 57.4 \times 10^{-4} \text{ J s}^{-1} \text{ K}^{-1} \\ &= 3.1 \times 10^{-4} \text{ J s}^{-1} \text{ K}^{-1} \end{aligned} \tag{3.41}$$

3.2.4 Entropy Flow due to IR Radiation from the Sky

Since IR radiation from the half of the sky hemisphere incident on a vertical surface is 0.18 cal cm^{-2} min^{-1} (Bartlett & Gates, 1967), the IR energy from the whole sky hemisphere incident on a horizontal plane is

$$\begin{aligned} E_{sky} &= 0.18 \times 2 \text{ cal cm}^{-2} \text{ min}^{-1} \\ &= 251 \text{ J m}^{-2} \text{ s}^{-1} \end{aligned} \tag{3.42}$$

Let us assume that the sky is a black body in the IR region. From the Stefan–Boltzmann Law, the effective temperature of the sky is given by

$$T_{sky} = \left[\frac{E_{sky}}{\sigma}\right]^{1/4} = 258.0 \text{ K} \tag{3.43}$$

where σ is the Stefan–Boltzmann constant. The entropy flux from the sky incident on a horizontal surface is (Planck, 1959, 1988)

$$\begin{aligned} \frac{4}{3}\sigma(T_{sky})^3 &= \frac{4}{3}\frac{E_{sky}}{T_{sky}} \\ &= 1.30 \text{ J m}^{-2} \text{ s}^{-1} \text{ K}^{-1} \end{aligned} \tag{3.44}$$

Hence, the entropy flux from the half of the sky hemisphere incident on a vertical surface becomes

$$\begin{aligned} S_{sky} &= \frac{1}{2} \times \frac{4}{3}\sigma(T_{sky})^3 \\ &= 0.649 \text{ J m}^{-2} \text{ s}^{-1} \text{ K}^{-1} \end{aligned} \tag{3.45}$$

The absorptivity of the lizard to IR radiation is $\alpha_t = 0.965$ (Bartlett & Gates, 1967). Thus the IR entropy from the sky absorbed by the lizard is

$$\begin{aligned} S^*_{\text{sky}} &= \alpha_t S_{\text{sky}} A_e \\ &= 0.965 \times 0.649 \times 57.4 \times 10^{-4}\ \text{J s}^{-1}\ \text{K}^{-1} \\ &= 36.0 \times 10^{-4}\ \text{J s}^{-1}\ \text{K}^{-1} \end{aligned} \tag{3.46}$$

3.2.5 Entropy Flow due to IR Radiation from the Ground

IR radiation from the half of the whole ground incident on a vertical surface is 0.32 cal cm^{-2} min^{-1} (Bartlett & Gates, 1967). Therefore, the IR energy from the whole ground incident on a horizontal plane is

$$\begin{aligned} E_{\text{grd}} &= 0.32 \times 2\ \text{cal cm}^{-2}\ \text{min}^{-1} \\ &= 447\ \text{J m}^{-2}\ \text{s}^{-1} \end{aligned} \tag{3.47}$$

Let us assume that the ground is a black body in the IR region. From the Stefan–Boltzmann Law, the effective temperature of the ground is given by

$$T_{\text{grd}} = \left(\frac{E_{\text{grd}}}{\sigma}\right)^{1/4} = 297.9\ \text{K} \tag{3.48}$$

The entropy flux from the ground incident on a horizontal surface is (Planck, 1959, 1988)

$$\begin{aligned} \frac{4}{3}\sigma(T_{\text{grd}})^3 &= \frac{4}{3}\frac{E_{\text{grd}}}{T_{\text{grd}}} \\ &= 2.00\ \text{J m}^{-2}\ \text{s}^{-1}\ \text{K}^{-1} \end{aligned} \tag{3.49}$$

Hence, the half of the entropy flux from the ground incident on a vertical surface becomes

$$\begin{aligned} S_{\text{grd}} &= \frac{1}{2} \times \frac{4}{3}\sigma(T_{\text{grd}})^3 \\ &= 0.999\ \text{J m}^{-2}\ \text{s}^{-1}\ \text{K}^{-1} \end{aligned} \tag{3.50}$$

Then, the IR entropy from the ground absorbed by the lizard is

$$\begin{aligned} S^*_{\text{grd}} &= \alpha_t S_{\text{grd}} A_e \\ &= 0.965 \times 0.999 \times 57.4 \times 10^{-4}\ \text{J s}^{-1}\ \text{K}^{-1} \\ &= 55.4 \times 10^{-4}\ \text{J s}^{-1}\ \text{K}^{-1} \end{aligned} \tag{3.51}$$

3.2.6 Entropy Flow due to IR Radiation Emitted by the Organism

Consider a mild situation with a lizard surface temperature of $T_{srf} = 34.1°C$ (Bartlett & Gates, 1967). The emissivity of the lizard ε for IR radiation is (Bartlett & Gates, 1967)

$$\begin{aligned} \varepsilon &= \alpha_t \\ &= 0.965 \end{aligned} \tag{3.52}$$

and the function $X(\varepsilon)$ becomes

$$X(0.965) = 1.01 \tag{3.53}$$

where $X(\varepsilon)$ is given by Eq. (3.27). The lizard can be regarded as a gray body with the temperature $T_{srf} = 34.1°C = 307.2$ K and with the emissivity $\varepsilon = 0.965$. The IR entropy flux emitted by the lizard as a gray body is given by [from Eq. (6.3) of Aoki, 1982a: the specific intensity $L = (1/\pi)(4/3)\varepsilon\sigma T^3 X(\varepsilon)$ and from the entropy $S = \int_0^{2\pi} \mathrm{d}\varphi \int_0^{\pi/2} L\cos\theta \sin\theta \, \mathrm{d}\theta = \pi L$]

$$\begin{aligned} S_{liz} &= \frac{4}{3}\varepsilon\sigma(T_{srf})^3 X(\varepsilon) \\ &= 2.14 \text{ J m}^{-2} \text{ s}^{-1} \text{ K}^{-1} \end{aligned} \tag{3.54}$$

Hence, the IR entropy emitted by the lizard becomes

$$\begin{aligned} S^*_{liz} &= S_{liz} A_e \\ &= 2.14 \times 57.4 \times 10^{-4} \text{ J s}^{-1} \text{ K}^{-1} \\ &= 122.6 \times 10^{-4} \text{ J s}^{-1} \text{ K}^{-1} \end{aligned} \tag{3.55}$$

3.2.7 Entropy Flow due to Convection

Consider again a mild situation with a lizard surface temperature of $T_{srf} = 34.1°C = 307.2$ K and with a wind speed of 0.447 m s^{-1}. This wind speed was a representative value for the location of the lizard near the ground (Bartlett & Gates, 1967). Under these conditions, the energy loss due to convection is (Bartlett & Gates, 1967)

$$\begin{aligned} E_{cnv} &= 18.6 \text{ cal min}^{-1} \\ &= 1.30 \text{ J s}^{-1} \end{aligned} \tag{3.56}$$

Hence, the entropy loss due to convection is

$$\begin{aligned} S^*_{cnv} &= \frac{E_{cnv}}{T_{srf}} \\ &= 42.2 \times 10^{-4}\ \mathrm{J\ s^{-1}\ K^{-1}} \end{aligned} \tag{3.57}$$

3.2.8 Entropy Flow due to the Evaporation of Water

The energy loss due to the evaporation of water is (Bartlett & Gates, 1967)

$$\begin{aligned} E_{evp} &= 0.2\ \mathrm{cal\ min^{-1}} \\ &= 0.0140\ \mathrm{J\ s^{-1}} \end{aligned} \tag{3.58}$$

Since the body temperature of the lizard is $T_{bd} = 34.1°\mathrm{C} = 307.2\ \mathrm{K}$ (Bartlett & Gates, 1967), the entropy loss due to evaporation becomes

$$\begin{aligned} S^*_{evp} &= \frac{E_{evp}}{T_{bd}} \\ &= 0.5 \times 10^{-4}\ \mathrm{J\ s^{-1}\ K^{-1}} \end{aligned} \tag{3.59}$$

3.2.9 Entropy Production due to Metabolic Heat Production

The metabolic heat production in the lizard is

$$\begin{aligned} E_{mtb} &= 0.6\ \mathrm{cal\ min^{-1}} \\ &= 0.0419\ \mathrm{J\ s^{-1}} \end{aligned} \tag{3.60}$$

and the body temperature of the lizard is $T_{bd} = 34.1°\mathrm{C} = 307.2\ \mathrm{K}$ (Bartlett & Gates, 1967). Hence, the entropy production due to metabolic heat production becomes

$$\begin{aligned} S^*_{mtb} &= \frac{E_{mtb}}{T_{bd}} \\ &= 1.4 \times 10^{-4}\ \mathrm{J\ s^{-1}\ K^{-1}} \end{aligned} \tag{3.61}$$

3.2.10 Entropy Flow due to Conduction

The incoming energy flows into the lizard thus far considered are due to direct solar radiation, diffuse solar radiation, reflected solar radiation, IR radiation from the sky, IR radiation from the ground, and metabolic heat production. The total amount of all this incoming energy is $43.2\ \mathrm{cal\ min^{-1}} = 3.01\ \mathrm{J\ s^{-1}}$ (Bartlett & Gates, 1967). However, the outgoing energy flows from the lizard thus far considered are due to IR radiation emitted by the lizard, convection, and the evaporation of water. The total amount of this outgoing energy is $57.7\ \mathrm{cal\ min^{-1}} = 4.03\ \mathrm{J\ s^{-1}}$ (Bartlett & Gates, 1967). If it is assumed that, on average, the incoming energy balances with the outgoing energy, then there should be another form of incoming energy of $14.5\ \mathrm{cal\ min^{-1}} = 1.01\ \mathrm{J\ s^{-1}}$. This amount of energy flow should be due to heat conduction to the lizard from a tree trunk:

$$\begin{aligned} E_{\mathrm{cnd}} &= 14.5\ \mathrm{cal\ min^{-1}} \\ &= 1.01\ \mathrm{J\ s^{-1}} \end{aligned} \tag{3.62}$$

Since the temperature of the lizard surface is $T_{\mathrm{srf}} = 34.1°\mathrm{C} = 307.2\ \mathrm{K}$, the entropy transported by heat conduction from a tree trunk to the lizard is

$$\begin{aligned} S^*_{\mathrm{cnd}} &= \frac{E_{\mathrm{cnd}}}{T_{\mathrm{srf}}} \\ &= 32.9 \times 10^{-4}\ \mathrm{J\ s^{-1}\ K^{-1}} \end{aligned} \tag{3.63}$$

3.2.11 Entropy Flow due to Mass-Flow Associated with Respiration

Resting oxygen consumption by lizards is given by (Bennett, 1983) $(1/5.6) \times 2.51 \times m^{0.78}\ \mathrm{mL\ O_2\ h^{-1}}$, where m is a body mass in grams. Since the body mass of the lizard that has been considered is $m = 18.4$ g (Bartlett & Gates, 1967), resting oxygen consumption becomes $4.35\ \mathrm{mL\ O_2\ h^{-1}} = 1.21 \times 10^{-3}\ \mathrm{mL\ O_2\ s^{-1}} = 5.39 \times 10^{-8}\ \mathrm{mole\ O_2\ s^{-1}}$. The CO_2 production is almost at the same level as the O_2 consumption within the range of experimental error (Nagi, 1983). The entropy content of O_2 gas in the standard conditions is $205.03\ \mathrm{J\ K^{-1}\ mole^{-1}}$, and that of CO_2 gas is $213.64\ \mathrm{J\ K^{-1}\ mole^{-1}}$; the difference between them is $8.61\ \mathrm{J\ K^{-1}\ mole^{-1}}$. Therefore, the entropy change due to the mass-flow associated with respiration is $5.39 \times 10^{-8} \times 8.61 = 4.6 \times 10^{-7}\ \mathrm{J\ s^{-1}\ K^{-1}}$, which is $10^{-2} - 10^{-4}$ of the amount of the entropy flows considered in the previous subsections and hence can be neglected here.

3.2.12 Total Entropy Flow and Entropy Production

From these calculations, the net sum of entropy flows into the lizard becomes

$$\begin{aligned} \Delta_e S &= (S^*_{\mathrm{dr}} + S^*_{\mathrm{sc}} + S^*_{\mathrm{rf}} + S^*_{\mathrm{sky}} + S^*_{\mathrm{grd}} + S^*_{\mathrm{cnd}}) - (S^*_{\mathrm{liz}} + S^*_{\mathrm{cnv}} + S^*_{\mathrm{evp}}) \\ &= -34.9 \times 10^{-4}\ \mathrm{J\ s^{-1}\ K^{-1}} \end{aligned} \tag{3.64}$$

Thus, the net entropy flow into the lizard is negative. That is, the lizard absorbs "negative entropy" (Schrödinger, 1944) from its surroundings. This negativity of the net entropy flow is the physical basis for maintaining the organized structures and functions in the lizard (Schrödinger, 1944). Net entropy flows from the lizard to the outside due to IR radiation, convection, and the evaporation of water are in the ratios 31.2 : 42.2 : 0.5 = 1.0 : 1.4 : 0.02.

If it is assumed that the lizard is in a steady state in entropy as in energy, then an entropy production $\Delta_i S$ should occur so as to result in no change of total entropy in the lizard

$$\begin{aligned}\Delta_i S &= -\Delta_e S \\ &= +34.9 \times 10^{-4}\ \mathrm{J\,s^{-1}\,K^{-1}}\end{aligned} \tag{3.65}$$

The positiveness of the entropy production $\Delta_i S$ shows (e.g., Chapter 1; Nicolis & Prigogine, 1977) that the Second Law of Thermodynamics certainly holds in the lizard as it does in a deer (Section 3.1). This is not self-evident without proof; its importance deserves special emphasis because even in the near past some authors (Beier, 1962; Trincher, 1967) erroneously asserted that the Second Law could not be applied to living organisms. The part of the entropy production $\Delta_i S$ that is due to metabolic heat production was already given and shown to be $S^*_{\mathrm{mtb}} = 1.4 \times 10^{-4}\ \mathrm{J\,s^{-1}\,K^{-1}}$. The remainder of the entropy production $33.5 \times 10^{-4}\ \mathrm{J\,s^{-1}\,K^{-1}}$ is due to irreversible biochemical reactions and motions of substances in the lizard not associated with metabolic heat production, as well as to irreversible heat flows in the lizard body and interaction with sunlight (reflection, scattering, absorption) within the body of lizard.

Since the body weight and the surface area of the lizard are, respectively, 18.4 g and 75.8 cm^2 (Bartlett & Gates, 1967), the entropy production per body weight is

$$\frac{34.9 \times 10^{-4}\ \mathrm{J\,s^{-1}\,K^{-1}}}{18.4\ \mathrm{g}} = 1.90 \times 10^{-4}\ \mathrm{J\,g^{-1}\,s^{-1}\,K^{-1}} \tag{3.66}$$

and that for per surface area is

$$\frac{34.9 \times 10^{-4}\ \mathrm{J\,s^{-1}\,K^{-1}}}{75.8\ \mathrm{cm^2}} = 0.460 \times 10^{-4}\ \mathrm{J\,cm^{-2}\,s^{-1}\,K^{-1}} \tag{3.67}$$

A pattern of entropy flow and entropy production in the lizard is shown in Figure 3.2. This flow-production diagram of entropy in the lizard summarizes and illustrates these results.

3.2.13 Comparison with White-Tailed Deer

In the previous section (Aoki, 1987d), the entropy production was calculated in a white-tailed deer during a winter night; the entropy production in a 50 kg female deer on a maintenance diet was 0.46 J s^{-1} K^{-1}. Thus, the entropy production of a lizard is 1/132 times smaller than the entropy production of a white-tailed deer. However, the entropy production per body weight for a lizard is 21 times larger than that for a deer, and the entropy production per effective surface area for a

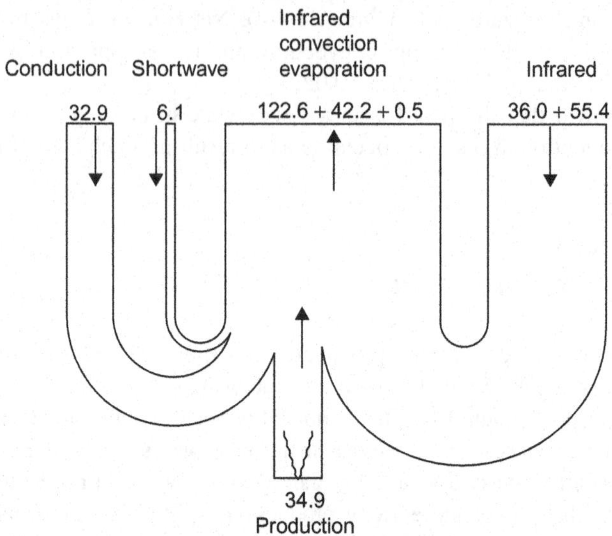

Figure 3.2 Flow and production diagram of entropy in a lizard in vertical position. Figures are in units of 10^{-4} J s^{-1} K^{-1}. Incoming shortwave solar entropy is 6.1 units, which consists of 1.4 units by direct solar radiation, 1.6 units by diffuse solar radiation, and 3.1 units by reflected solar radiation. Incoming IR entropy consists of 36.0 units from the sky and 55.4 units from the ground. Entropy inflow by conduction is 32.9 units. Entropy outflows due to IR radiation, convection and evaporation are 122.6 units, 42.2 units, and 0.5 units, respectively. The entropy of 34.9 units is produced within a lizard.

lizard is 1.4 times that of a deer and comparable. These trends are similar to the case of the BMR in thermal physiology (Kleiber, 1975).

3.3 Pigs

3.3.1 Newborn Pigs

Butchbaker and Shanklin (1964) measured various modes of heat loss from a newborn pig suspended in a calorimeter chamber (direct calorimetry—see the next chapter). The results for a newborn pig [age = 1 day, weight = 3 lb (1.36 kg)] are shown in their figure 7, giving total heat loss, sensitive heat loss, radiation energy loss, convection heat loss, and latent heat loss at air temperatures of 55−100°F (12.8−40.6°C). Using their results, the corresponding entropy productions in a newborn pig are calculated in the following way (Aoki, 1992).

Radiation energy loss is expressed as (the Stefan−Boltzmann law)

$$R = \sigma \varepsilon A^{e}[(T_{s})^{4} - (T_{e})^{4}] \tag{3.68}$$

where σ is the Stefan–Boltzmann constant, ε is the emissivity of pig surface for IR radiation (equal to the absorptivity according to the Kirchhoff's law), A^e is the effective radiating surface area of the pig, T_s is the surface temperature of the pig, and T_e is the ambient temperature. The emissivity (ε) is taken as $\varepsilon = 0.92$ (Bond, Kelly, and Heitman, 1952), and A^e is assumed to be $0.75A$ (Bond et al., 1952), where $A = 0.097$ (body weight in kg)$^{0.633}$ is the total surface area of the pig in units of m^2 (Deighton, 1932). As an example, let us consider the case of the ambient (air) temperature of $T_e = 80°F$ (26.7°C, 299.8 K). In this case, $R = 4.88\ J\ s^{-1}$ from Butchbaker and Shanklin (1964; figure 7). Then, from Eq. (3.68), the surface temperature of the pig is $T_s = 309.2$ K.

The entropy flow accompanying outgoing IR radiation from the pig is [for radiation entropy, refer to, Eq. (3.54) or, e.g., Aoki, 1982a, 1987b]

$$S_{out} = \frac{4}{3}\sigma\varepsilon A^e (T_s)^3 X(\varepsilon = 0.92) \tag{3.69}$$

where

$$X(\varepsilon) = \frac{45}{4\pi^4}\frac{1}{\varepsilon}\int_0^\infty y^2[(x+1)\ln(x+1) - x\ln x]dy \qquad x = \frac{\varepsilon}{e^y - 1} \tag{3.70}$$

The factor $X(\varepsilon)$, introduced by Landsberg and Tonge (1979), has often been used in the previous sections. For $T_s = 309.2$ K,

$$S_{out} = 0.185\ J\ s^{-1}\ K^{-1} \tag{3.71}$$

The entropy flow of IR radiation absorbed by the surface of the pig is likewise given by

$$S_{in} = \frac{4}{3}\sigma\varepsilon A^e (T_e)^3 \tag{3.72}$$

where it is assumed that the IR emissivity of the inside wall of the calorimeter in which the pig is suspended is equal to 1 [$X(1) = 1$; see Landsberg & Tonge (1979)]. (Note that the absorptivity equals $\varepsilon = 0.92$ according to the Kirchhoff's law.) At $T_e = 299.8$ K,

$$S_{in} = 0.166\ J\ s^{-1}\ K^{-1} \tag{3.73}$$

The other modes of heat exchange between the pig and its environment are the conduction–convection of heat and the evaporation heat of water. The sum of these E^* is obtained as the total heat loss (E_{tot}) minus the radiation energy loss (R), and the value E_{tot} is plotted in Butchbaker and Shanklin (1964; figure 7): $E_{tot} = 8.21$ (J s^{-1}). Hence, $E^* = E_{tot} - R = 3.32$ (J s^{-1}). The entropy flow due to conduction–convection and evaporation becomes

$$S^* = \frac{E^*}{T_s} = 0.011 \text{ J s}^{-1} \text{ K}^{-1} \tag{3.74}$$

Thus, the net entropy flow into the pig is

$$S_{flow} = S_{in} - S_{out} - S^* = -0.030 \text{ J s}^{-1} \text{ K}^{-1} \tag{3.75}$$

Entropy flow S_{flow}, entropy production S_{prod}, and the change of entropy content ΔS are related as (Chapter 1; Nicolis & Prigogine, 1977)

$$\Delta S = S_{flow} + S_{prod} \tag{3.76}$$

Let us assume that the entropy in the pig is at steady state and is kept almost constant; that is, $\Delta S \simeq 0$. This assumption is reasonable because the deep body temperature in even a newborn pig is kept nearly constant, and the change with time of the heat content (and hence the change of the entropy content) of the pig is very small (Mount, 1959, 1964). Hence, the entropy production in the pig and per surface area are

$$S_{prod} = -S_{flow} = 0.030 \text{ J s}^{-1} \text{ K}^{-1} \tag{3.77}$$

and

$$\frac{S_{prod}}{A} = 0.258 \text{ J m}^{-2} \text{ s}^{-1} \text{ K}^{-1} \tag{3.78}$$

Similar calculations are carried out for other ambient temperatures by using the data in Butchbaker and Shanklin (1964; figure 7). Figure 3.3 shows the results S_{prod}/A plotted against the total heat (energy) loss per surface area E_{tot}/A. As shown, the entropy production per surface area is a linear function of the total heat loss per surface area. Figure 3.3 also includes the results of similar calculations based on another energetic study on a newborn pig (weight = 2 kg) made by Mount (1964), showing the same tendency.

Figure 3.3 shows that for newborn pigs the lower the ambient temperature is, the larger the total heat loss and the entropy production become. This is due to a rise in the metabolic rate of a newborn pig that is reacting to exposure to cool environments to maintain homeothermy (Mount, 1959; Mount & Rowell, 1960a). The rise in metabolism induces the increase of heat production and hence the increase of heat loss and of entropy production.

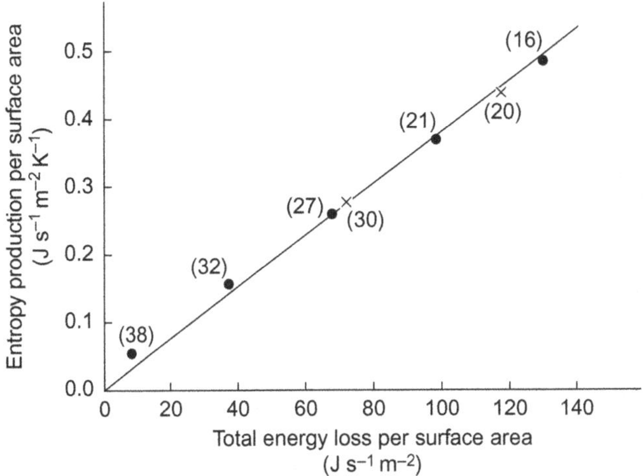

Figure 3.3 Entropy production per surface area S_{prod}/A versus total heat (energy) loss per surface area E_{tot}/A for newborn pigs. Values in parentheses are ambient temperatures in °C. ●: Based on Butchbaker and Shanklin (1964); ×: based on Mount (1964).

3.3.2 Mature Pigs

Bond et al. (1952) investigated heat loss from mature pigs in a calorimeter chamber. In their study, pigs in groups of four or five were brought to the calorimeter chamber, which was kept as similar as possible to actual farm conditions. Three groups (A, B, and C) were used for measuring heat loss.

- *Group A*: 26−37 kg in ambient temperatures 40−100°F (4−37°C),
- *Group B*: 34−57 kg in 50−100°F (9−37°C),
- *Group C*: 125−173 kg in 40−100°F (4−37°C).

Due to the pigs' habit of huddling, the exposed surface area differs from the total surface area; the degree of huddling and hence the exposed area depends on the ambient temperature. Bond et al. (1952) estimated the average exposed areas for pigs in each group at various ambient temperatures. The results of the measurement of heat loss of an average pig in each group are shown in Bond et al. (1952; figures 4−6).

Based on their data, similar calculations as in Section 3.3.1 are carried out (Aoki, 1992), and the results are shown in Figure 3.4. Also, in this case, the entropy production per surface area changes linearly with the total heat loss per surface area.

As in newborn pigs, lower ambient temperatures give larger heat losses and entropy productions, although the correlation in this case is not as clear as in newborn pigs.

As shown, group C (the heaviest and oldest) have rather smaller values of total heat loss and of entropy production than those of groups A and B at similar ranges of ambient temperature.

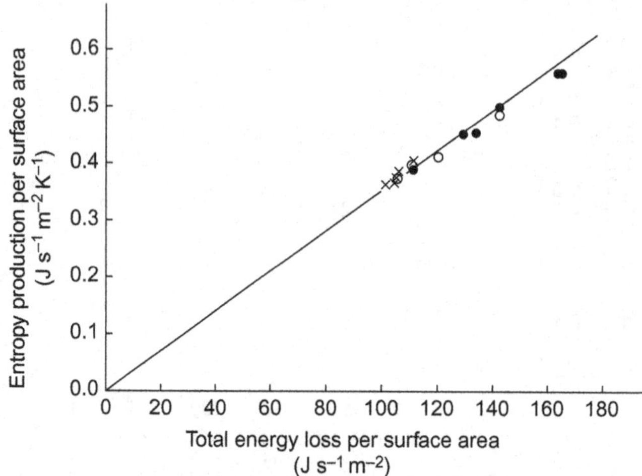

Figure 3.4 Entropy production per surface area S_{prod}/A versus total heat (energy) loss per surface area E_{tot}/A for mature pigs. ●: Group A; ○: group B; ×: group C. The explanation for each group is given in the text.

The values of total heat loss and of entropy production for mature pigs fall into narrower ranges than those in newborn pigs (shown in Figure 3.3), indicating that abilities for keeping constant physiological conditions within the body are well developed in matures pigs. That is, mature pigs are less sensitive to the change of environmental conditions than newborns.

Slopes of the straight lines in Figures 3.3 and 3.4 are almost the same; every point in both figures is nearly on the same straight line of the slope of $0.0035-0.0036\ K^{-1}$.

3.3.3 Age Dependence

Mount (1959) showed that even newborn pigs with little hair and subcutaneous fat (low insulation against heat loss) exhibit a vigorous metabolic response to the cooling of the environment and that the rectal temperature (a representative of the deep body temperature) of pigs is kept almost constant. It is suggested that thermogenesis in pigs is well developed at birth (Mount & Rowell, 1960a, 1960b). Therefore, the change of heat content in pigs will be very small, and heat loss can be considered to equal heat production (Mount, 1964).

Since the entropy production per surface area is proportional to the total heat loss per surface area, as shown previously, the entropy production per surface area is thus proportional to the heat production per surface area.

Brody and Kibler (1944) found that the heat production per surface area for pigs increases from birth to a maximum at about 7 months of age and tends to gradually decrease thereafter. Since the heat production per surface area is proportional to the

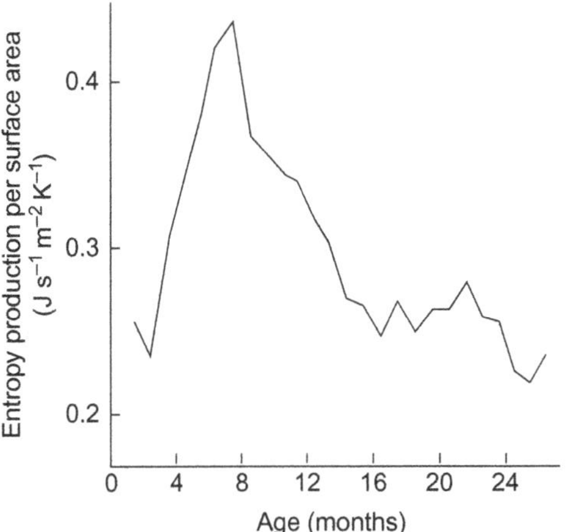

Figure 3.5 Entropy production per surface area versus age for pigs (based on the energetic study of Brody & Kibler, 1944).

entropy production per surface area, as previously stated, it is concluded that the entropy production per surface area for pigs increases from birth until about 7 months of age and then gradually decreases. That is, pigs show two stages of the initial increase and the later decrease for entropy production profile per surface area. Figure 3.5 shows these trends (based on the energetic data of Brody & Kibler, 1944).

3.3.4 Discussion

This section is an attempt to support the so-called multistages hypothesis (initial increase and later decrease of entropy production), which is different from the Prigogine–Wiame principle (Chapter 1, note 2). The Prigogine–Wiame principle is proved to be theoretically valid only for systems near equilibrium (Nicolis & Prigogine, 1977). For systems far from equilibrium, any corresponding alternative principles have not been obtained. Biological organisms are evidently in states far from equilibrium. Hence the only way to obtain entropy laws for organisms will be to calculate entropy production using observed energetic data and to infer laws governing organisms, as is done in this section. The result is the multistages hypothesis.

Entropy and entropy-related quantities are the only physical quantities that have directionality with time (Aoki, 1991). Hence the investigation of directionality is always important for the entropy-related phenomena concerned. This section has shown that two- or multidirectionality is preferred in the life span of pigs, not the one-directionality of the Prigogine–Wiame principle. Generally speaking, the activity of life starts from nearly "zero" level, then grows to a "nonzero" level, and finally returns to "zero" level at death. The two- or multistages hypothesis (initial increase and later decrease) is well in accord with this intuitive image about life span (this is not the case for the Prigogine–Wiame principle). In this respect, the multistages hypothesis will reflect the essential trend of the life span of organisms. This is consistent with plants (Figure 2.4) and will also be shown in the later chapters, dealing with humans and even ecological systems.

4 Humans I: Direct Calorimetry

The energy characteristics of a human body as a whole have been measured by direct and indirect calorimetry. In direct calorimetry, a human subject is put in a closed calorimetric chamber, and heat loss and in some cases various modes of energy-flow are measured. In indirect calorimetry, oxygen absorption by humans is measured; then from the specific relation between heat production and the amount of absorbed oxygen, heat production in humans is estimated. For details of calorimetry the reader is referred to, for example, Kleiber (1975) and Kemp (1999).

From the both methods, the energetic natures of humans are known in detail, and, by the use of them, the entropy productions of humans can be calculated, as in this and next chapters (Aoki, 1989c, 1990b, 1991, 1994).

4.1 Direct Calorimetry: Outline and Significance

Energy inflow into a naked human body in a respiration calorimeter is due to IR radiation emitted by the inside walls of the calorimeter, and energy outflow from the human body is due to IR radiation emitted by the body, convection in its surrounding air, and the evaporation of water from the lung and the body surface. Hardy and DuBois (1938a, 1938b) measured heat exchange for the human body through radiation, convection, and evaporation by the use of the Hardy radiometer and the respiration calorimeter, in which a nude subject is laid in basal conditions. They also measured mass-flows into and out of the human body.

From these results, it is possible, as shown in this chapter, to calculate the corresponding entropy inflows and outflows and to estimate the entropy production occurring within the human body. The concept of entropy has not been treated in human physiology, but entropy is as important a concept as energy, as thermodynamics shows: the First Law of Thermodynamics is concerned with the concept of energy, and the Second and Third Laws with entropy. As shown in Chapter 1, entropy production is a kind of global measure that specifies the strength of motions and reactions occurring in nature. Hence, entropy production in the human body shows the extent of activity of (1) heat flows and (2) the motions and reactions of substances within the body as a whole. So entropy production is a significant physiological quantity that characterizes the human body from the thermodynamic and holistic (i.e., considering a human body as a whole) viewpoints.

Entropy Principle for the Development of Complex Biotic Systems. DOI: 10.1016/B978-0-12-391493-4.00004-4

4.2 Calculation of Entropy Production

In the classical experiment by Hardy and DuBois (1938a), the naked subject (EFDB) is put in the respiration calorimeter in basal conditions for 1 h. The subject emits IR radiation to its surroundings. The average skin temperature of the subject is $T_s = (1/2)(33.02 + 32.91) + 273.15 = 306.12\ \mathrm{K}$ when the calorimeter temperature is 27.40°C (Hardy & DuBois, 1938a). The emissivity of the human skin for IR radiation is nearly 1 (Hardy, 1934). Therefore, the entropy flow associated with IR radiation emitted by the subject is (for radiation entropy, for example, refer to Aoki, 1982a, 1983, 1998; Planck, 1959, 1988)

$$\begin{aligned} S_{\mathrm{out}} &= 1.54 \times \frac{4}{3}\sigma(T_s)^3 \\ &= 3.340\ \mathrm{J\,s^{-1}\,K^{-1}} \end{aligned} \tag{4.1}$$

where $\sigma = 5.670 \times 10^{-8}\ \mathrm{J\,m^{-2}\,s^{-1}\,K^{-4}}$ is the Stefan–Boltzmann constant, and 1.54 is the effective radiating surface area of the subject (EFDB) in units of m^2 (Hardy & DuBois, 1938a).

However, the subject in the calorimeter absorbs IR radiation emitted by the inside walls of the calorimeter. The temperature of the calorimeter is $T_c = 27.40°\mathrm{C} = 300.55\ \mathrm{K}$ (Hardy & DuBois, 1938a). The absorptivity of the human skin to IR radiation is equal to the IR emissivity according to the Kirchhoff's law and is nearly 1. Therefore, the entropy flow associated with IR radiation emitted by the inside walls of the calorimeter and absorbed by the subject is

$$\begin{aligned} S_{\mathrm{in}} &= 1.54 \times \frac{4}{3}\sigma(T_c)^3 \\ &= 3.161\ \mathrm{J\,s^{-1}\,K^{-1}} \end{aligned} \tag{4.2}$$

The energy lost by convection is (Hardy & DuBois, 1938a)

$$\begin{aligned} E_{\mathrm{cnv}} &= 11.00\ \mathrm{kcal\,h^{-1}} \\ &= 12.78\ \mathrm{J\,s^{-1}} \end{aligned}$$

Hence, the corresponding entropy lost by convection is

$$\begin{aligned} S_{\mathrm{cnv}} &= E_{\mathrm{cnv}}/T_s \\ &= 0.042\ \mathrm{J\,s^{-1}\,K^{-1}} \end{aligned} \tag{4.3}$$

Similarly, the energy lost by the evaporation of water is (Hardy & DuBois, 1938a)

$$\begin{aligned} E_{\mathrm{evp}} &= 20.66\ \mathrm{kcal\,h^{-1}} \\ &= 24.01\ \mathrm{J\,s^{-1}} \end{aligned}$$

Assume that the lung temperature is same as the rectal temperature (identified with the body temperature). The rectal temperature is $T_r = 37.1°C = 310.3$ K when the calorimeter temperature is 27.40°C (estimated from figure 1 of Hardy & DuBois, 1938b). Then, the entropy lost by evaporation becomes

$$\begin{aligned} S_{evp} &= E_{evp}/T_r \\ &= 0.077 \text{ J s}^{-1} \text{ K}^{-1} \end{aligned} \tag{4.4}$$

Thus, the net entropy flow into the subject due to energy-exchange is given by

$$\begin{aligned} S_{flow}(\text{energy}) &= S_{in} - (S_{out} + S_{cnv} + S_{evp}) \\ &= -0.298 \text{ J s}^{-1} \text{ K}^{-1} \end{aligned} \tag{4.5}$$

Next, consider entropy flows associated with mass-flows into and out of the subject. The consumed O_2 content in respiration is 20.67 g h^{-1} = 1.794 × 10^{-4} mol s^{-1} (Hardy & DuBois, 1938a), and the entropy content of O_2 gas at the standard conditions (298.15 K and 1 atm) is 205.03 J K^{-1} mol^{-1}. Hence, the entropy flow by O_2 uptake is $S(O_2) = 1.794 \times 10^{-4} \times 205.03 = 0.0368$ J s^{-1} K^{-1}. Likewise, the CO_2 gas liberated by the subject is 23.16 g h^{-1} = 1.462 × 10^{-4} mol s^{-1} (Hardy & DuBois, 1938a), and the entropy content of CO_2 gas is 213.64 J K^{-1} mol^{-1}. Hence, the entropy of the emitted CO_2 gas is $S(CO_2) = 1.462 \times 10^{-4} \times 213.64 = 0.0312$ J s^{-1} K^{-1}. The net entropy into the subject associated with respiration is thus $S(O_2) - S(CO_2) = 0.006$ J s^{-1} K^{-1}; this is very small compared with S_{flow} (energy) = −0.298 J s^{-1} K^{-1} [about 1.9% of S_{flow}(energy)].

The H_2O liberated from the subject is 35.42 g h^{-1} = 5.466 × 10^{-4} mol s^{-1} (Hardy & DuBois, 1938a) and the entropy content of H_2O gas is 188.72 J K^{-1} mol^{-1}. The entropy outflow associated with H_2O is 5.466 × 10^{-4} × 188.72 = 0.103 J s^{-1} K^{-1}. It consists of two parts: (1) the entropy associated with the evaporation heat of water that has already been taken into account by Eq. (4.4) ($S_{evp} = 0.077$ J s^{-1} K^{-1}) and (2) the remaining $S(H_2O) = 0.026$ J s^{-1} K^{-1} that is the entropy content of liquid water supplied from the body of the subject to the outside. The entropy flow due to mass-exchange is thus

$$\begin{aligned} S_{flow}(\text{mass}) &= S(O_2) - S(CO_2) - S(H_2O) \\ &= -0.020 \text{ J s}^{-1} \text{ K}^{-1} \end{aligned} \tag{4.6}$$

The total entropy flow into the subject becomes from Eqs (4.5) and (4.6)

$$\begin{aligned} S_{flow} &= S_{flow}(\text{energy}) + S_{flow}(\text{mass}) \\ &= -0.318 \text{ J s}^{-1} \text{ K}^{-1} \end{aligned} \tag{4.7}$$

Thus, the net entropy flow into the human body is negative. This means that the human body absorbs "negative entropy" from its surroundings, as Schrödinger

(1944) asserted, as in plants (Chapter 2) and in animals (Chapter 3). This is the physical basis for maintaining the ordered structures and functions in the human body (Schrödinger, 1944).

The energy does not balance in this 1 h experiment done by Hardy and DuBois (1938a). The heat eliminated from the body is 77.36 kcal h^{-1}, and the heat produced is 68.65 kcal h^{-1} (Hardy & DuBois, 1938a); that is, the change of heat content of the body is $\Delta Q = -8.71\ \text{kcal h}^{-1} = -10.1\ \text{J s}^{-1}$ (the increase of heat is considered positive and the decrease negative). The corresponding entropy change in the body becomes

$$\Delta S(\text{heat}) = \Delta Q/T_r = -0.033\ \text{J s}^{-1}\ \text{K}^{-1} \tag{4.8}$$

where $T_r = 310.3$ K is the rectal temperature (corresponding to the body temperature) at the calorimeter temperature of 27.40°C, which is taken from figure 1 of Hardy and DuBois (1938b).

Also, water does not balance in this experiment: the water in the body is consumed and liberated to the outside. The entropy of the water consumed and liberated to the outside has already been obtained as

$$\Delta S(\text{water}) = -S(\text{H}_2\text{O}) = -0.026\ \text{J s}^{-1}\ \text{K}^{-1} \tag{4.9}$$

Thus, the total entropy change in the body becomes

$$\begin{aligned}\Delta S &= \Delta S(\text{heat}) + \Delta S(\text{water})\\ &= -0.059\ \text{J s}^{-1}\ \text{K}^{-1}\end{aligned} \tag{4.10}$$

On the average of one day, water is supplied to the body from the outside, and the entropy change due to the change of water content in the body will be near zero: $\Delta S(\text{water}) \simeq 0$. Similarly, $\Delta S(\text{heat}) \simeq 0$, and hence $\Delta S \simeq 0$ on the average of one day. That is, the body is at steady state in entropy in the longer term (over the course of a day).

The entropy change (ΔS) of the body is the sum of two terms: the entropy flow into the body (S_{flow}) and the entropy production within the body (S_{prod}). That is, $\Delta S = S_{\text{flow}} + S_{\text{prod}}$ (e.g., Chapter 1; Nicolis & Prigogine, 1977). The entropy flow is already given in Eq. (4.7), and the entropy change of the body is given in Eq. (4.10). Then, the entropy production $S_{\text{prod}} = \Delta S - S_{\text{flow}}$ for the subject EFDB is given by

$$S_{\text{prod}} = 0.259\ \text{J s}^{-1}\ \text{K}^{-1} \tag{4.11}$$

The entropy production per effective radiating surface area is

$$S_{\text{prod}}/1.54\ \text{m}^2 = 0.168\ \text{J m}^{-2}\ \text{s}^{-1}\ \text{K}^{-1} \tag{4.12}$$

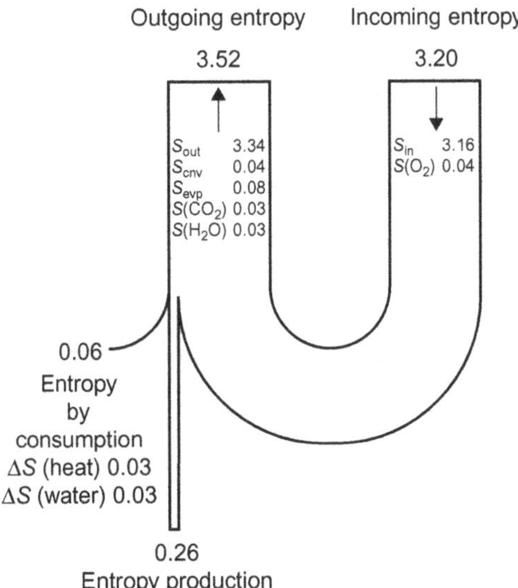

Figure 4.1 Flow-production diagram of entropy in the human body (the subject EFDB) in units of J s^{-1} K^{-1} at surrounding temperature of 27.4°C. The incoming entropy of 3.20 units consists of 3.16 units by IR radiation absorbed by the human body and 0.04 units by oxygen uptake. The entropy of 0.26 units is produced within the body. The entropy by consumption of 0.06 units consists of 0.03 units by heat consumption and 0.03 units by water consumption. The outgoing entropy of 3.52 units consists of 3.34 units by emission of IR radiation, 0.04 units by convection, 0.08 units by evaporation of water, 0.03 units by CO_2 liberation, and 0.03 units by liquid water supplied to the outside.

and that for per body weight is

$$S_{prod}/74.74\ \text{kg} = 0.00347\ \text{J kg}^{-1}\ \text{s}^{-1}\ \text{K}^{-1} \tag{4.13}$$

That is, the entropy production S_{prod} is positive. This means that the Second Law of Thermodynamics is valid in the human body, as is evident from the formulation of the Second Law in irreversible thermodynamics for open systems (Chapter 1; Nicolis & Prigogine, 1977). This is against the arguments given by Beier (1962) and Trincher (1967) that the Second Law cannot be applied to living organisms. The flow-production diagram of entropy for the human body is made from these results and shown in Figure 4.1.

As just shown, the entropy flow associated with mass-flow in respiration [$S(O_2)$ − $S(CO_2)$] is small compared with the net entropy flow associated with energy-exchange [S_{flow} (energy)] and is less than the last significant figure. Neglecting this term [$S(O_2) - S(CO_2)$] results in the increase of the value of entropy production (S_{prod}) by only 2.3%. Hence, it will be neglected in the following discussions.

Also, the term $S(H_2O)$ is contained in both S_{flow} and ΔS [see Eqs (4.6), (4.7), (4.9), and (4.10)] and canceled in the calculation of entropy production by the equation $S_{prod} = \Delta S - S_{flow}$. Hence, to obtain entropy production, the term $S(H_2O)$ does not need to be taken into account.

4.3 Entropy Homeostasis

The calorimeter temperature dependence of human entropy production is examined by methods similar to those described using the data of Hardy and Dubois (1938b).

The results are shown in Figure 4.2. Entropy production is constant from 26°C to 32°C (the thermally neutral zone) of calorimeter temperature (below 25°C, see Section 4.4.3), and the average is 0.172 J m^{-2} s^{-1} K^{-1} for the two subjects (EFDB and JDH).

Also, surrounding air currents from the fan and clothing in the calorimetric chamber have only small (increasing) effects on human entropy production (from the data of Hardy, Milhorat, & DuBois, 1938a).

This approximate constancy to the environment in basal conditions may be called the "entropy homeostasis" of humans in basal conditions.

4.4 Nonbasal Effects

Nonbasal effects (exercise and chills) on entropy production are investigated based on the energetic studies of Hardy, Milhorat, and DuBois (1938b) and DuBois (1939), using methods similar to those shown in Section 4.2.

4.4.1 *Mild Exercise*

In Hardy et al. (1938b, table 2), the effect of mild exercise on the energetics of nude subjects (EFDB and JDH) in a respiration calorimeter is shown. Some of these results are also shown diagrammatically in Hardy et al. (1938b, figure 1) and are used for the calculation of human entropies. The results show that entropy production increases 50–140% by mild exercise due to the increase in

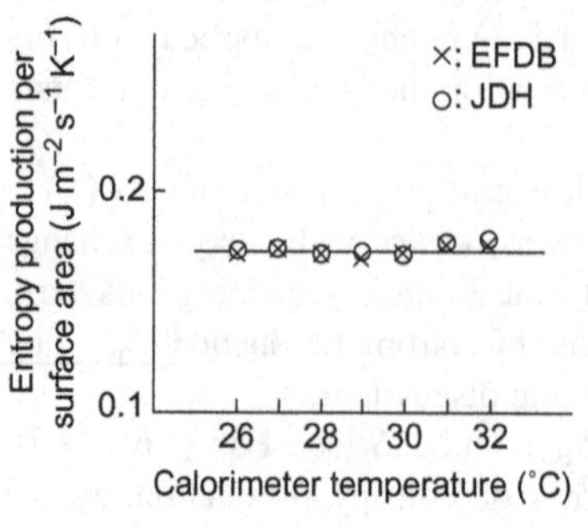

Figure 4.2 Entropy production per effective radiating surface area for the subjects EFDB and JDH in units of J m^{-2} s^{-1} K^{-1} at calorimeter temperatures of 26–32°C. The average is 0.172, and the standard deviation is 0.003.

heat production within the body by exercise (Hardy et al., 1938b). In the quiet period of 1 h after exercise, the entropy productions drop nearly to the basal levels.

4.4.2 Violent Exercise

The energetic data on more violent exercise (a game of squash) are shown in Hardy et al. (1938b, figure 3), although the values are only approximate. From these data, it is possible to obtain the entropy production within the body after violent exercise by using a similar method to the one just described; the results are shown in Figure 4.3.

The entropy production per effective radiating surface area after violent exercise is six to eight times as great as that before exercise. Comparison with the previous results shows that the entropy production increases as the exercise becomes more violent. In the rest after exercise, the entropy production decreases, gradually approaching to the basal level.

4.4.3 Chills in Cold Environments

When a nude subject lies at below 25°C for an hour or so in the calorimetric chamber, the subject begins to feel chilly, and then involuntary violent contractions of the muscles and shivering of the whole body occur. The energetics observed during this chill period is shown in Hardy et al. (1938b, table 1), and some of the data are also diagrammatically shown in Hardy et al. (1938b, figure 1, the first and the second diagrams). In the chill period, the heat regulation mechanism is called upon, and much heat is produced within the body (Hardy et al., 1938b). Hence, the entropy produced should be mostly associated with this heat production during the chill.

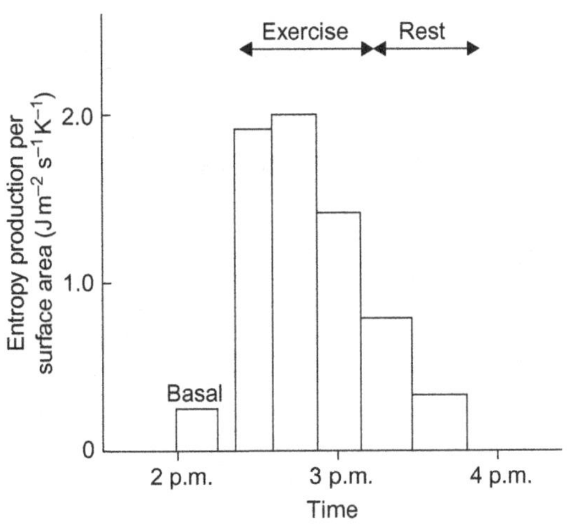

Figure 4.3 The effect of violent exercise on the entropy production per effective radiating surface area of the human body. The first period is in basal condition, the next three periods are in violent exercise, and the last two periods are in rest.

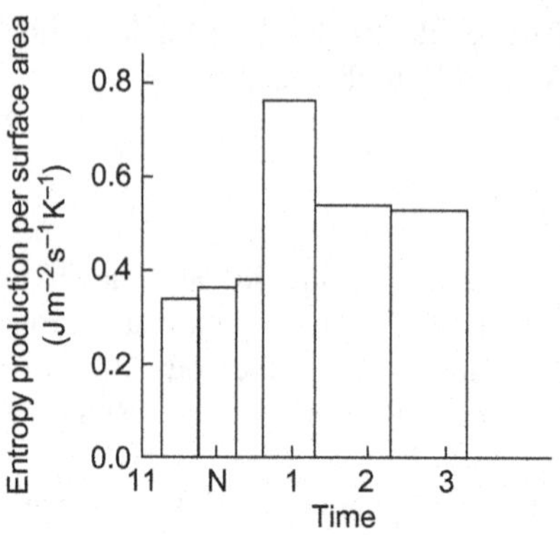

Figure 4.4 The effect of a malarial chill on the entropy production per effective radiating surface area of the human body. The first three periods are before a malarial chill, the fourth period is during a chill, and the last two periods are after a chill.

The entropy production during the chill can be calculated by using a method similar to the one just described. The results show that the entropy production in the chill period is about twice as large as that in basal conditions without the chill. In the quiet period after the chill, the entropy production falls near to the basal value. Thus, the chill is a kind of entropy-producing as well as heat-producing mechanism. The extent of entropy production during the chill is of the same order of magnitude as that in mild exercise.

4.4.4 Malarial Chill

Chills that occur in disease with a resultant fever were also studied. Barr and DuBois (1918) made calorimeter experiments on patients in different periods of malarial fever. The third diagram of DuBois (1939, figure 10) shows the results of Barr and DuBois (1918) on the patient George S., combined with observations made later on the other patients. From these data, entropy productions in the patient George S. in malarial fever were calculated by using a method similar to the one just described [1.83 m^2 was employed for the total surface area of George S., according to Barr and DuBois (1918)]. The results are shown in Figure 4.4. The entropy productions before a malarial chill (the first three periods) are nearly at the level of 0.36 $J\ m^{-2}\ s^{-1}\ K^{-1}$, which is about 2.1 times the entropy production in normal subjects in basal conditions. During a malarial chill (the fourth period), the entropy production rises to 0.76 $J\ m^{-2}\ s^{-1}\ K^{-1}$, about 4.4 times that of normal subjects. After a malarial chill (the last two periods), the entropy production falls to 0.53 $J\ m^{-2}\ s^{-1}\ K^{-1}$, about 3.1 times that of normal subjects.

Thus, a malarial chill produces much entropy within the human body, as in the case of chills in cold environments. Moreover, even before a chill, the patient in malarial fever retains a high level of entropy production; after a chill, the entropy production falls but does not reach the level before a chill within even 2 h. These points (i.e., a high level in entropy production before a chill and higher after a chill) are contrasted to the case of chills in cold environments discussed previously. These are the characteristics of patients in malarial fever and perhaps in other fevers.

5 Humans II: Indirect Calorimetry

Indirect calorimetry is the measurement of the amount of oxygen absorbed by a subject through respiration. Based on the specific relation of oxygen absorption and heat production, heat production in the subject is estimated. In the human body, 1 L of oxygen consumption corresponds to 4.82 kcal (20.2 kJ) of heat production. When the subject is in an energetic steady state, heat production equals heat loss, and indirect calorimetry and direct calorimetry yield the same results.

Heat production is equal to the so-called dissipation function in physical chemistry. The dissipation function, divided by the absolute temperature of an organism, gives the entropy production within the organism (Section 5.2).

5.1 Biological Meanings

Biotic activity within most organisms is supported by oxygen uptake (respiration). Incorporated oxygen is used to decompose macromolecules in the body, such as carbohydrate, protein, and lipid; high-quality chemical energy is liberated. This chemical energy is employed for miscellaneous ordered chemical reactions and for the motions of matter that are necessary to keep the biotic system in "lively state." This energy production corresponds to the dissipation function. Chemical energy thus used gradually deteriorates, finally becomes low-quality heat energy, and is discarded by organisms to the outside as heat loss.

To maintain homeostatic ordered structure and function, the entropy content in organisms should be retained at a constant low level because entropy is a measure of randomness. In biotic systems, many highly organized networks of physical, chemical, and organic reactions and motions are needed to sustain biological order within the system. These reactions and motions are essentially irreversible; they are inevitably entropy producing according to the Second Law of Thermodynamics, and the entropy content of the system tends to increase. Hence, to maintain a steady level of entropy content in the system (orderliness), the excess entropy (randomness) that is produced within the system should always be discarded to the outside, along with heat energy (the dissipation function). Discarded entropy production is the dissipation function divided by the temperature of the biotic system (Section 5.2).

Entropy Principle for the Development of Complex Biotic Systems. DOI: 10.1016/B978-0-12-391493-4.00005-6

5.2 Physical Formulation

5.2.1 Dissipation Function

Entropy production in physicochemical systems is expressed by (Haase, 1969, p. 85)

$$\frac{\mathrm{d_i}S}{\mathrm{d}t} = \frac{1}{T}\frac{\mathrm{d}W_{\mathrm{diss}}}{\mathrm{d}t} + \frac{1}{T}\sum_{\rho} A_{\rho}\frac{\mathrm{d}\xi_{\rho}}{\mathrm{d}t} \tag{5.1}$$

where W_{diss} is the work associated with dissipative effects, A_{ρ} is the affinity, and $\mathrm{d}\xi_{\rho}/\mathrm{d}t$ is the rate of chemical reaction ρ. Dissipation function is defined as (Haase, 1969, p. 85)

$$\psi = T\frac{\mathrm{d_i}S}{\mathrm{d}t} = \frac{\mathrm{d}W_{\mathrm{diss}}}{\mathrm{d}t} + \sum_{\rho} A_{\rho}\frac{\mathrm{d}\xi_{\rho}}{\mathrm{d}t} \tag{5.2}$$

Entropy production and the dissipation function are integral indicators of the strength of all physical and chemical processes in the system.

When p and T are constant, it is shown that (Haase, 1969, p. 41)

$$A_{\rho} = -\left(\frac{\partial G}{\partial \xi_{\rho}}\right)_{T,p} \tag{5.3}$$

where G is the Gibbs function: $G = H - TS$. Then, the second term of Eq. (5.2) is expressed as

$$\sum_{\rho} A_{\rho}\frac{\mathrm{d}\xi_{\rho}}{\mathrm{d}t} = \sum_{\rho}\left[-\left(\frac{\partial H}{\partial \xi_{\rho}}\right)_{T,p} + T\left(\frac{\partial S}{\partial \xi_{\rho}}\right)_{T,p}\right]\frac{\mathrm{d}\xi_{\rho}}{\mathrm{d}t} \tag{5.4}$$

In usual cases, it is possible to neglect the second term in the right-hand side (Prigogine, 1967, p. 27), that is,

$$\sum_{\rho} A_{\rho}\frac{\mathrm{d}\xi_{\rho}}{\mathrm{d}t} \approx \sum_{\rho} -\left(\frac{\partial H}{\partial \xi_{\rho}}\right)_{T,p}\frac{\mathrm{d}\xi_{\rho}}{\mathrm{d}t} \tag{5.5}$$

Hence, the second term of ψ [Eq. (5.2)] becomes the rate of heat production in chemical reactions (Prigogine, 1967, p. 10):

$$\sum_{\rho} A_{\rho}\frac{\mathrm{d}\xi_{\rho}}{\mathrm{d}t} \approx \sum_{\rho} {r_{T,p}}^{(\rho)}\frac{\mathrm{d}\xi_{\rho}}{\mathrm{d}t} \tag{5.6}$$

where ${r_{T,p}}^{(\rho)}$ is the heat of reaction associated with a change of ξ_{ρ} at constant T and p.

Dissipated work W_{diss} is finally converted to heat energy in the system by dissipative processes.

Thus, the dissipation function ψ is the rate of production of heat in the system.

5.2.2 Energy Metabolism in Organisms

In the energy metabolism of most living systems, all processes are powered by oxygen consumption and glycolysis. Hence, Zotin (1990, p. 59) expressed that

$$\psi = \frac{dq(O_2)}{dt} + \frac{dq(gl)}{dt} \tag{5.7}$$

where the first term is energy production by oxygen consumption, and the second term is energy production by glycolysis. For most organisms under aerobic conditions, the rate of glycolysis is very small compared with the oxygen consumption rate, and it can be neglected. Then, ψ can be expressed as (Zotin, 1990, p. 59)

$$\psi = \frac{dq(O_2)}{dt} \tag{5.8}$$

Thus, the dissipation function (ψ) is equal to the rate of the metabolic energy production by oxygen consumption in living systems. Energy production by oxygen consumption ψ becomes finally heat energy, and it should be discarded to the outside as heat loss, maintaining the homeostasis (steady state) of organisms.

5.2.3 Psiufunction

It may happen that small amounts of the dissipation function do not go out from the organisms in special conditions, remaining fixed inside living systems. Zotin (1990, p. 206) gave the name "psiufunction" to these quantities and denoted them as ψ_u. It corresponds to the neglected term in chemical heat production [Eq. (5.6)] and a part of the dissipated work dW_{diss}/dt. However, the external dissipation function $dq(out)/dt$ was called "psidfunction" (ψ_d) (Zotin, 1990, p. 206). That is,

$$\psi = \psi_d + \psi_u, \quad \psi_u = \psi - \psi_d = \frac{dq(O_2)}{dt} - \frac{dq(out)}{dt} \tag{5.9}$$

It turns out that ψ_u is almost zero in adult organisms in ordinary conditions (Zotin, 1990). Hence,

$$\psi_u = 0, \quad \psi = \frac{dq(O_2)}{dt} = \frac{dq(out)}{dt} \tag{5.10}$$

However, ψ_u is not zero in certain special conditions of living systems, such as embryonic development, work, starvation, animal hibernation, plant dehydration,

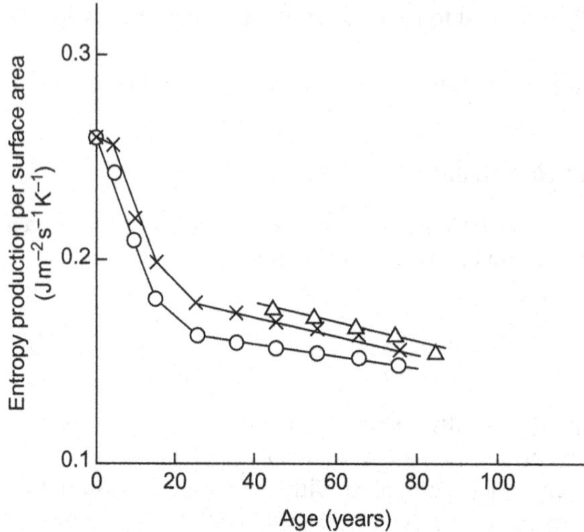

Figure 5.1 Entropy production per effective radiating surface area versus 0–85 years of age for men and women. ×: Japanese male; o: Japanese female; △: male (Shock, 1955).

abnormal temperature, different substances and diseases, and so on (Zotin, 1990). But these cases are not considered here.

5.3 Age Dependence I: Per Surface Area

Indirect calorimetry is simple to handle, and entropy production is easy to calculate. Dissipation functions (metabolic energy productions $\psi \equiv E_{mtb}$) per surface area at various ages can be observed. Here they are taken from Sasaki (1979, table 2) for the Japanese male and female and from Shock (1955, table I) for males. Also, the average rectal temperature T_r (body temperature) is taken from DuBois, Ebaugh, and Hardy (1952, figure 16): $T_r = 310.2$ K for the male and $T_r = 310.1$ K for the female. They are assumed to be applicable to all ages.

Entropy productions per surface area at various ages (E_{mtb}/T_r) are shown in Figure 5.1 (0–85 years) and Figure 5.2 (0–10 years).

As shown, the entropy production increases during 0–2 years of age. In this range of age, the difference in entropy production between men and women is not significant. From 2 to 25 years of age, the entropy production decreases rapidly and then decreases gradually to 75–85 years. The difference between men and women becomes significant in this range of age. The difference between Japanese males and the males in the study by Shock (1955) is due to the difference of race or food. Even at extreme ages (75–85 years of age), entropy production seems to be decreasing, and a constancy or a level has not been achieved.

5.4 Age Dependence II: Per Individual

This section deals with entropy production in the human individual. Entropy productions in the individual at various ages are obtained from entropy productions

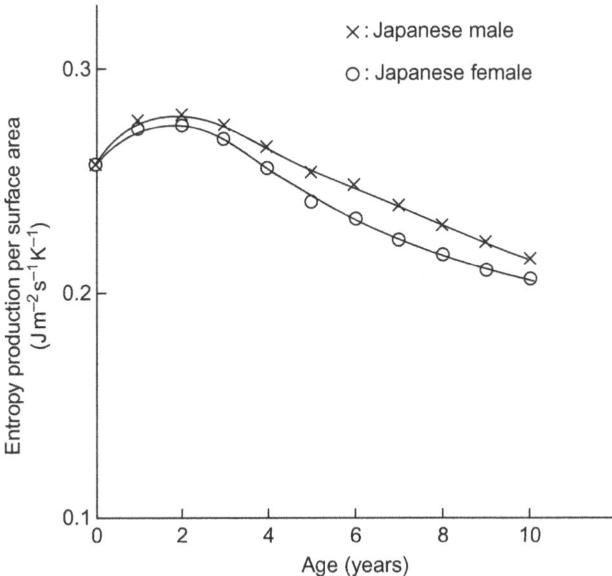

Figure 5.2 Entropy production per effective radiating surface area versus 0–10 years of age for men and women. The key is the same as for Figure 5.1.

per surface area (Section 5.3) multiplied by the average surface area given by, for example, Sasaki (1985, table 3.3) which are known from weight and height at different ages according to the empirical formula.

The results are shown in Figure 5.3 (0–75 years) and Figure 5.4 (0–19 years). Entropy production increases from birth to about 16 years of age for males and to about 14 years of age for females, then gradually decreases afterward. Entropy production in the human body is a kind of measure of the activity of motion and of the interaction of energy and matter occurring within the body, as shown in Chapter 1. Hence, the results obtained here are intuitively recognized as reasonable. Close views of the behavior of entropy production before the maximums, such as rapid increases just after birth, can be seen in Figure 5.4.

Thus, entropy production in an individual manifests two phases over the human life span: the initial increasing stage and the later decreasing stage. These trends have been observed in the case of per surface area (Section 5.3) but are demonstrated more clearly and without any doubt in the present case (Figures 5.3 and 5.4).

5.4.1 The Average Human Individual: Extrapolation

Averages of entropy production for the male and female at various ages are adopted as the entropy production of a standard human being and plotted (based on the data of Sasaki, 1985, table 3.3) and extrapolated in Figure 5.5.

The solid line is based on experimental observations (ages 0–75 years). The maximum occurs at about 15 years, which is considered the most active stage in

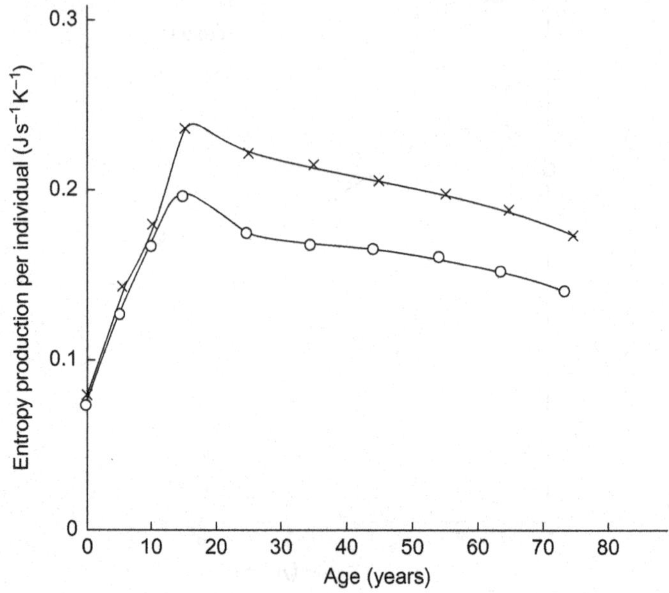

Figure 5.3 Entropy production in the individual versus 0–75 years of age for male and female humans. ×: male; ○: female.

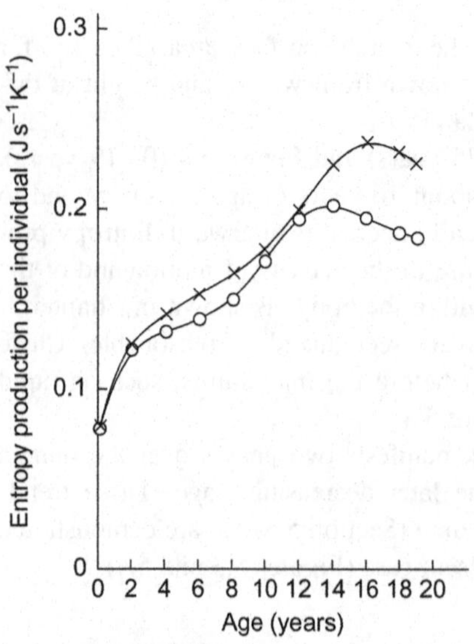

Figure 5.4 Entropy production in the individual versus 0–19 years of age for male and female humans. ×: male; ○: female.

life. It is common sense in medical sciences that aging occurs as early as in the twenties. Human life begins from a fertilized egg (about 9 months before birth). This is shown as point A in Figure 5.5, obtained by extrapolation of the solid line. The entropy production of a fertilized egg is $0.06\ \mathrm{J\ s^{-1}\ K^{-1}}$. Entropy

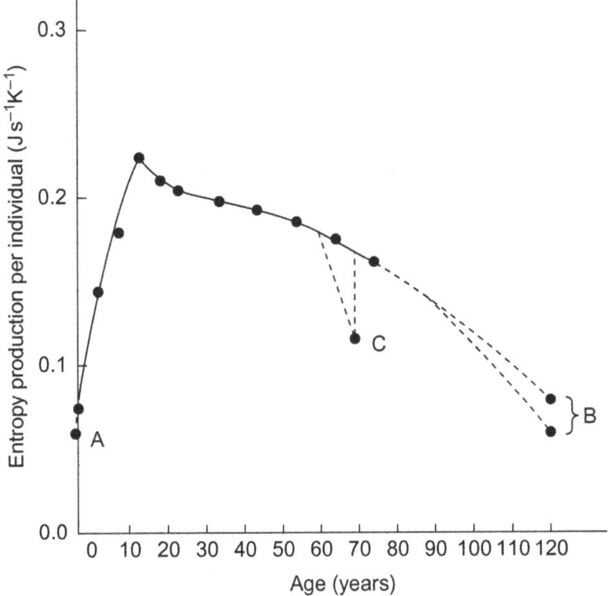

Figure 5.5 Entropy production in a standard human being versus age. Explanations are given in the text.

production increases rapidly from the fertilized egg stage to 15 years of age and decreases slowly thereafter until 75 years of age. No data are available beyond 75 years; hence the solid line is rather arbitrarily extrapolated to 120 years, the maximal human life span considered, perhaps, as genetically determined. Entropy production at 120 years may be shown by point B in Figure 5.5: 0.06–0.08 J s^{-1} K^{-1}. Note that points A and B are at about the same level. This value of entropy production, 0.06–0.08 J s^{-1} K^{-1}, may represent the minimal value for the human individual to exist in a living state.

Point C in Figure 5.5 represents death by disease or accident as an example. The position of C differs depending on circumstances. Entropy production does not become zero immediately upon death because some organs are still alive even at death. This is the medical basis for organ transplants. After death, entropy production immediately becomes zero by cremation but gradually by interment.

5.4.2 A Principle of Organization

As shown in Figure 5.5, entropy production rises rapidly in the initial period of the human life span from a fertilized egg to 15 years of age and decreases gradually afterward. In other words, the strength of activity in the human body shows a rapid increase in the initial stage and a gradual decrease thereafter.

It is probable that this trend can be generalized to the life spans of other organisms. During the initial growing stage, the system is incomplete, fragile, and vulnerable to external perturbations. Hence, to build a solid organized structure, the initial phase should take place in a short period of time by concentrating a large amount of activity. Otherwise, organisms could not survive and could not become mature individuals. In the context of Darwin's principle of natural selection, present-day organisms would be the result of a selection mechanism for survival during the long period of evolution process. After completing their structure and function, organisms become solid, and a long aging process begins with a slow decrease of activity.

Thus, it is proposed that a Principle of Organization in complex biotic systems is constituted of a rapid increase of activity in an initial short period of development, followed by a slow decrease of activity in a long period of aging. This principle may be applied to the organization processes of other systems.

6 Ecological Communities

An ecological community is the aggregate of groups of various species in ecological systems. Organisms in the community eat one another and are eaten by one another, and they make up a trophic structure consisting of trophic compartments (food chain and food web). Many diagrams of food chains and food webs have been presented in ecology (see Figure 6.3 as an example). Here, properties of entropy production in aquatic communities (lakes and estuaries) are investigated in some detail, which lead to the Min–Max Principle of entropy production for the development of ecological communities (Aoki, 2006).

6.1 Respiration

As already shown in Section 5.1, biotic activities in most organisms are supported by oxygen uptake, that is, respiration. Associated with respiration, high-quality energy is produced within organisms and used for miscellaneous chemical, physical, and organic reactions to keep biotic systems in an ordered and lively state. This energy is the dissipation function. [For example, in the human body, 1 L of absorbed oxygen (respiration) creates 4.82 kcal of heat—dissipation function.] This energy gradually deteriorates, becomes low-quality heat energy, and is discarded to the outside as heat loss.

These activities in biotic systems are essentially irreversible and necessarily produce entropy according to the Second Law of Thermodynamics. To keep homeostatic structure and function (order) in systems, the entropy produced should also be discarded to the outside together with dissipation function. Entropy production is the dissipation function divided by the absolute temperature of living systems (Section 5.2).

Respiration, the dissipation function, and entropy production are thus closely related, although they are measured in different physical units. Usually, respiration is measured either as oxygen uptake per unit time (e.g., L/h) by the excretion of carbon dioxide, or by the corresponding biomass from organisms, which is determined as the difference between digested food and production (Browder, 1993). The associated dissipation function is measured in energy units (e.g., kcal or kJ per hour). For the method of measurement of respiration in aquatic systems, the reader is referred to Lampert (1984).

Entropy Principle for the Development of Complex Biotic Systems. DOI: 10.1016/B978-0-12-391493-4.00006-8

6.2 Trophic Diversity

Let a biomass of *i*th trophic compartments in a food web be denoted as B_i and $B = \sum_i B_i$, where summation is over all the biotic compartments (except for detritus). (For trophic compartments in the food web, see, e.g., Figure 6.3.) For trophic diversity D of the food web, Simpson's diversity index (1949) is adopted, as in Aoki and Mizushima (2001) and Aoki (2003):

$$D = \frac{1}{\sum_i p_i^2} \tag{6.1}$$

where $p_i = B_i/B$. This diversity index takes into account both the distribution pattern of the biomass of trophic compartments and the richness of the trophic compartments in a whole ecosystem. Note that it differs from species richness. In the present study, the view is shifted from species to trophic compartment (functional group). This diversity index is better and more convenient than species richness in describing the diversity of large and complex ecological systems simply from a whole-systemic viewpoint.

6.3 Trophic Diversity Versus Respiration

6.3.1 Three Estuaries in the USA

Monaco and Ulanowicz (1997) made a comparative study of three mid-Atlantic estuaries on the eastern U.S. coast. Each estuary is composed of 13–14 trophic compartments. The biomass of each trophic compartment, the biomass flow among trophic compartments, the flow to detritus, the export/import to/from outside, and the respiration of each trophic compartment are specified. The units of biomass are mg C m^{-2}, and the units of flow are mg C m^{-2} $year^{-1}$. From these values, total biomass B, total respiration R, and trophic diversity D for each estuary can be obtained. Plots of R/B versus D are given in Figure 6.1 for the three estuaries (cross symbols): from left to right in the figure, the estuaries of Chesapeake Bay, Narragansett Bay, and Delaware Bay.

6.3.2 Three Japanese Lakes

The IBP/PF project in Japan in 1966–1972 investigated the productivity of communities of Japanese inland waters, including five lakes, two rivers, and one culture pond (as well as one tropical lake in Malaysia) (Mori & Yamamoto, 1975). For the five lakes, the biomass of phytoplankton, macrophytes, planktonic bacteria, zooplankton, zoobenthos, and fish were measured in g C m^{-2}. Community respiration was estimated in units of g C m^{-2} $year^{-1}$ for three lakes: Lake Yunoko, Lake Suwa, and Lake Kojima. From these values, community respiration per total

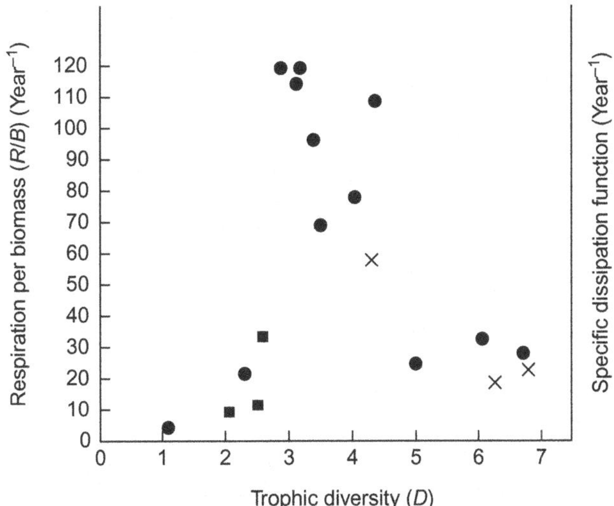

Figure 6.1 Respiration per biomass versus trophic diversity in aquatic communities. ■: Lake Kojima, Yunoko, and Suwa. ×: Estuaries of Chesapeake Bay, Narragansett Bay, and Delaware Bay. ●: Broa Reservoir, Veli Lake, Lake Malawi (1993), Turkana, Ontario, Tanganyika (1980), Tanganyika (1974), Kinneret, Malawi (1995), George, Victoria (1971), and Victoria (1985).

biomass R/B and trophic diversity D can be obtained for these lakes. R/B versus D data are plotted in Figure 6.1 for these three lakes (solid squares): Lake Kojima, Lake Yunoko, and Lake Suwa from left to right in the figure.

6.3.3 ECOPATH Project

The ECOPATH method, developed by Christensen and Pauly (1992, 1993, http://www.ecopath.org), was constructed for the quantification of a steady-state multi-species aquatic system. It provides the biomass of each trophic compartment, a quantitative pattern of biomass flow among trophic compartments, the flow to detritus, the export/import to/from the outside, and the respiration in each compartment (except primary producers). The lack of respiration in primary producers in ECOPATH causes its results to differ from those of other network analysis methods (e.g., NETWRK) (Heymans & Baird, 2000).

Because the respirations of phytoplankton and aquatic plants are not given in the ECOPATH model, let us try here to make a model calculation to estimate them. Figure 6.2 shows a compartment of phytoplankton (or aquatic plants). Gross production a in phytoplankton is consumed by production p, flow to detritus d, and respiration r. Thus,

$$a = r + p + d \tag{6.2}$$

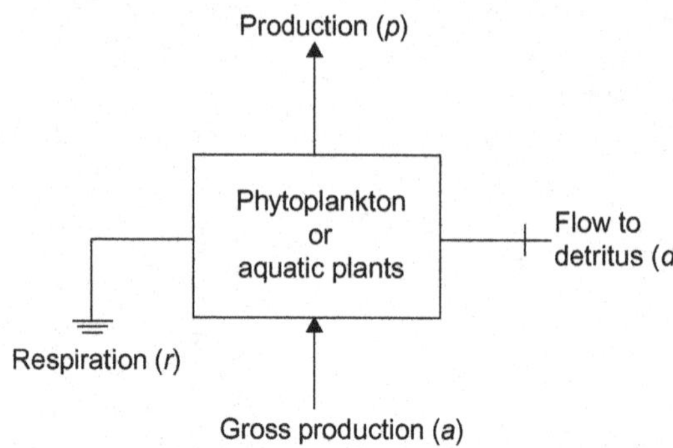

Figure 6.2 A trophic compartment of phytoplankton or aquatic plants.

Here, let us assume that

$$r = \alpha a \tag{6.3}$$

From Eqs (6.2) and (6.3)

$$r = (p + d)\frac{\alpha}{1 - \alpha} \tag{6.4}$$

Thus, r can be estimated from values of p, d, and α. The values of p and d are indicated in each ECOPATH diagram; hence the value α is needed to estimate r.

6.3.3.1 Phytoplankton

The study of Riley (1946) gave α of 0.42 per year for Georges Bank. Hogetsu, Kitazawa, Kurasawa, Shiraishi, and Ichimura (1952) showed the average annual value of $\alpha = 0.47$ in Lake Suwa (from a citation in Aruga, 1973). In the IBP/PF project in Japan, the annual respiration–gross production ratio α was given as the average value of 0.44 for Lake Biwa, Lake Yunoko, Lake Suwa, and Lake Kojima (Mori & Yamamoto, 1975). Monaco and Ulanowicz (1997) showed $\alpha = 0.42$ on average for the estuaries of Narragansett Bay, Delaware Bay, and Chesapeake Bay. For the purpose of a model calculation, the average value 0.4 is adopted for α of phytoplankton, in accordance with the well-accepted assumption of Steemann Nielsen (1960).

6.3.3.2 Aquatic Plants

Data on respiration and gross production in aquatic plants are fewer than those of terrestrial communities and phytoplankton. Ikusima (1966) studied the monthly average of tape grass (*Vallisneria denseserrulata*) in its growing period (June–October) and gave the value of respiration–gross production ratio α of 0.64. This grass begins to grow in late April and dies in early October. Hence the ratio $\alpha = 0.64$ may be regarded as an approximate annual value for this grass. In a similar manner, the data of Kurasawa, Tezuka, Kobori, and Aoyama (1962) on the aquatic plant *Hydrilla verticillata* (growing in May–December) provide the annual value 0.65 for α. Odum (1957) gave an annual value of $\alpha = 0.57$ for the spring community of *Sagittaria lorata*. Tanimizu and Miura (1976) gave the annual average $\alpha = 0.67$ in the submerged plant *Egeria densa*. From these values, 0.65 is adopted as an annual value of α in the present model calculation for aquatic plants.

6.3.3.3 An Example: Lake Turkana

Let us consider Lake Turkana (1987–model) as an example. There is no respiration in phytoplankton in the original ECOPATH flow diagram (Kolding, 1993, figure 3), which is also reproduced in figure 18.25 in Begon, Harper, and Townsend (1996). All flows are in g m^{-2} year^{-1}. In the phytoplankton box, production is $p = 1{,}300$, and the flow to detritus is $d = 4{,}266.3$. Then, using $\alpha = 0.4$, Eq. (6.4) gives $r = 3{,}710.9$. The improved food web diagram is shown in Figure 6.3. Total respiration R in the food web becomes 6,213.5, and total biomass $B = 54.8$. Hence, respiration per biomass for the Lake Turkana food web becomes $R/B = 6{,}213.5/54.8 = 113.4$. Trophic diversity is calculated as $D = 3.1$ from Eq. (6.1). In Figure 6.1, Lake Turkana is plotted as the fourth solid circle from the left of the figure.

6.3.3.4 Other Lakes

By means of similar procedures, the results of other lakes in ECOPATH diagrams are plotted as solid circles in Figure 6.1: Broa Reservoir (Angelini & Petrere, 1996), Veli Lake (Aravindan, 1993), Lake Malawi (Degnbol, 1993), Lake Ontario (Christensen, 1995), Lake Tanganyika (1980, 1974) (Moreau, Nyakageni, Pearce, & Petit, 1993), Lake Kinneret (Walline, Pisanty, Gophen, & Berman 1993), Lake Malawi (Allison, Patterson, Irvine, Thompson, & Menz, 1995), Lake George (Moreau, Christensen, & Pauly, 1993), and Lake Victoria (1971, 1985) (Moreau, Ligtvoet, & Palomares, 1993).

6.4 Eutrophication

Consider a relationship, if any, between trophic diversity and the degree of eutrophication of aquatic systems. Figures 6.4 and 6.5 give the relationship between trophic diversity and some measures of eutrophication: chlorophyll-a and the total phosphorus concentration at the water surface (International Lake Environment Committee, 1994).

The values for the degree of eutrophication are average ones, and fluctuations around average values are somewhat large within months, seasons, and years in some aquatic systems. Nevertheless, the average tendency in Figures 6.4 and 6.5 may suggest a positive correlation between trophic diversity and the degree of eutrophication. This relationship is assumed here as a general trend in the present aquatic systems concerned, although there will be exceptions in general (the opposite is discussed later).

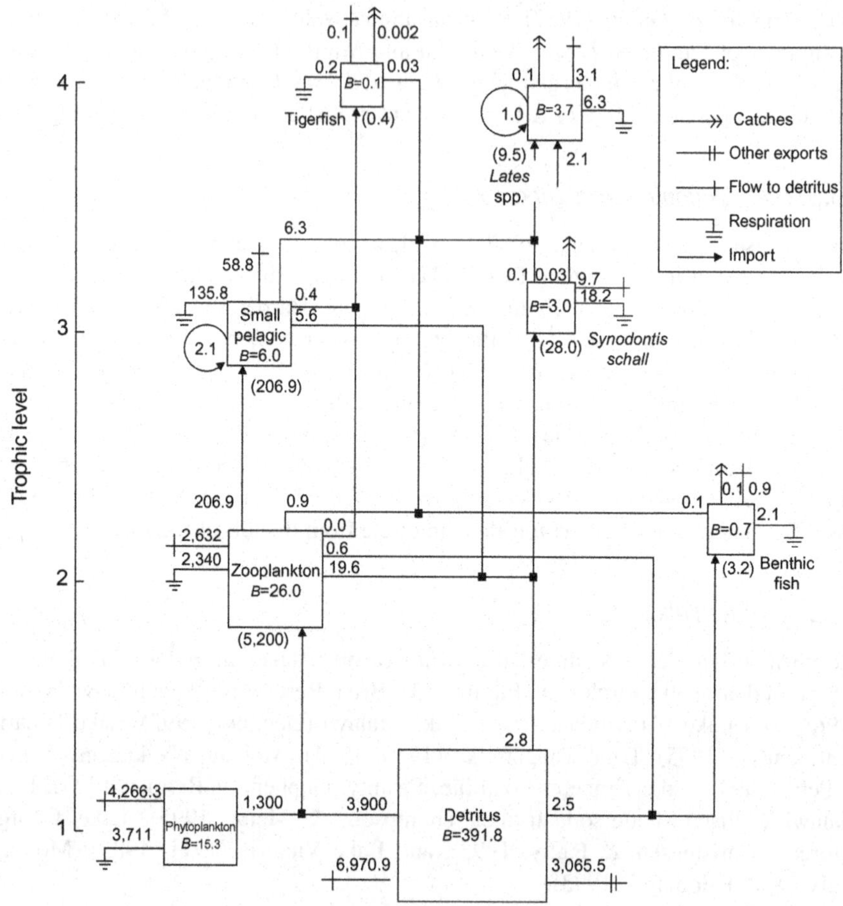

Figure 6.3 ECOPATH food web diagram for Lake Turkana. The respiration of phytoplankton is added. Biomasses are in g m^{-2}, and flows are in g m^{-2} year^{-1}. The B in each box means B_i here.
Source: Modified from figure 3 of Kolding (1993).

6.5 Arrow of Time

Eutrophication is a slow and natural process in the geological history of a lake (Lampert & Sommer, 1997), though it has been accelerated by anthropogenic activities in recent years. Thus, the degree of eutrophication is a so-called arrow of time in aquatic ecosystems. Nature consists of many hierarchical structures, each of which has its own characteristic concepts and laws (Anderson, 1972), and each hierarchy has its own "time" (Saito, 2002). Eutrophication can be regarded as a characteristic time in a hierarchy of aquatic ecosystems. Thus, Figure 6.1 shows

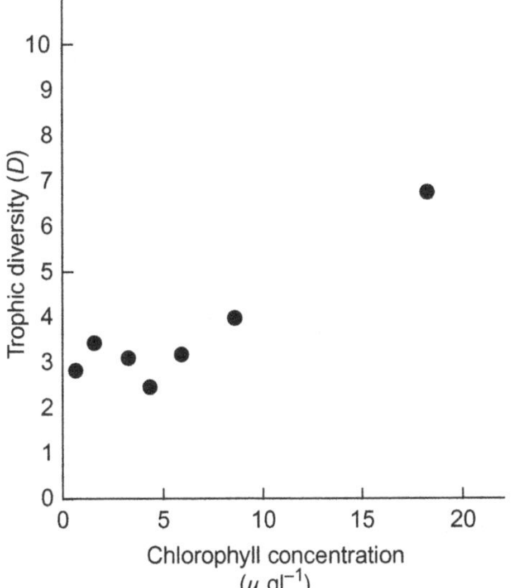

Figure 6.4 Trophic diversity versus chlorophyll-a concentration. From left to right: Lake Malawi, Tanganyika (north). Turkana, Biwa (northern), Ontario, Kinneret, and Victoria (1985).

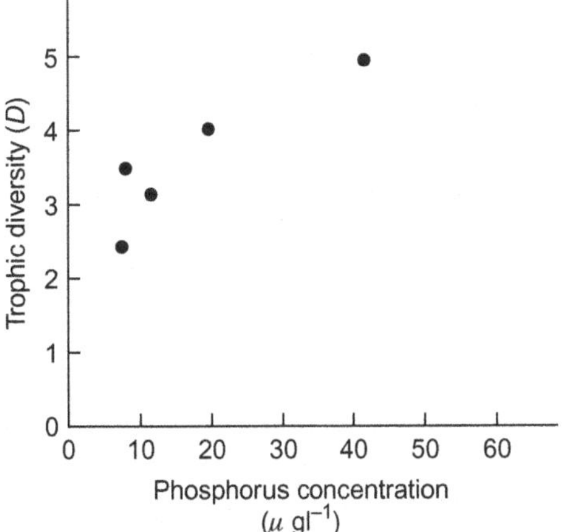

Figure 6.5 Trophic diversity versus total phosphorus concentration. From left to right: Lake Biwa (lower symbol), Tanganyika (upper symbol). Ontario, Kinneret, and George.

that respiration per biomass consists of two phases: an initial increase and a later decrease with respect to eutrophication and hence with time.

6.6 The Dissipation Function and Entropy Production

Section 6.1 shows that liberated respiration energy is equal to dissipation function. Thus, the specific dissipation function (dissipation function per biomass) is equal to respiration per biomass (Figure 6.1). The specific dissipation function of aquatic

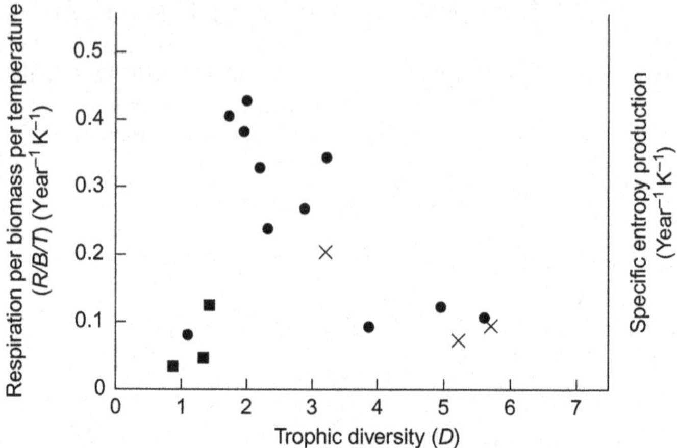

Figure 6.6 Specific entropy production versus trophic diversity in aquatic communities. Symbols and their order are the same as in Figure 6.1, except that Broa Reservoir is omitted because data on water temperature are not available.

communities has the two-phase (up-and-down) character with respect to diversity, eutrophication, and time.

Organisms in lakes and estuaries are poikilotherms, and their temperature is nearly equal to the temperature of the environment, that is, the temperature (absolute) of the surrounding water. Hence, the dissipation function divided by the absolute temperature of the water is regarded as equal to the entropy production of communities. The difference of temperature—for example, $\Delta T = 10°C$ of average water—leads to only a small change of difference on the order of $\Delta T/273 = 3.7\%$ or less. Hence, the general tendency of the dissipation function is also applied to entropy production.

Specific entropy production (respiration per biomass per absolute temperature of water) versus trophic diversity is shown in Figure 6.6. Water surface temperatures (the annual average) are taken from Mori and Yamamoto (1975), the Japan Oceanographic Data Center (private communication), and the International Lake Environment Committee (1994). As in the dissipation function, entropy production has the two-phase (up-and-down) characteristic with respect to diversity, eutrophication, and essentially with time. The schematic profile is shown in Figure 6.7. The time scale cannot be specified: it may be of geological order.

Also, entropy production in organisms [pigs (Chapter 3) and humans (Chapter 5)] increases initially and decreases later. Thus, this two-directional characteristic may be a universal property in biotic systems, including organisms and ecological systems. It may be characterized as a Biological Principle, in contrast to the Physical Principle (a variation principle or principle of least action), which is always one-directional (Aoki, 1989a).

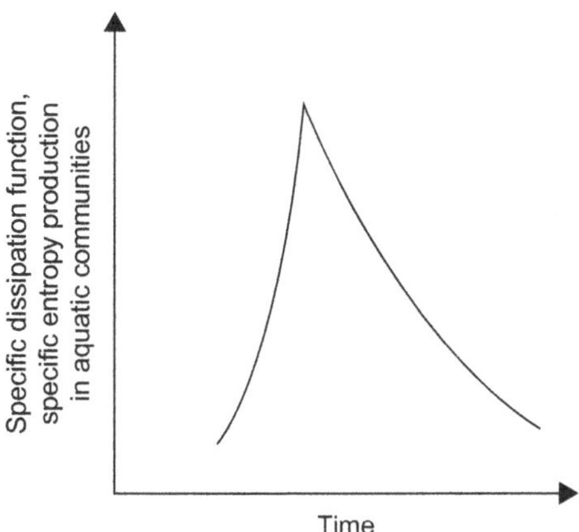

Figure 6.7 Schematic profile of the specific dissipation function and specific entropy production in aquatic communities. Scales are arbitrary. The time scale may be of geological order.

6.7 Discussion

The present chapter is concerned with some ordinary lakes and estuaries, and the line of discussion may not be applicable to other specific aquatic ecosystems. For example, the relationship between diversity and eutrophication may be reversed in some aquatic system, such as coral reefs, where oligotrophication may lead to high diversity. Then, in Figure 6.1, the direction of the abscissa of eutrophication and hence of time is inverted. Nevertheless, the two-phase characteristic is not changed by this time-reversal, as seen in Figure 6.1 (the initial phase becomes the later phase and vice versa), and the view taken in this chapter does not change. Specific aquatic ecosystems such as coral reefs, upwelling systems, plankton blooms, and so on should be considered separately and are left for future study.

7 Lake Ecosystems

An ecosystem is a system consisting of organisms (communities) and their physical environments. This chapter deals with lake ecosystems. As pointed out by Hutchinson (1964), the study of large and complex ecosystems, such as lakes, consists of two different approaches: holological (holos = whole) and merological (meros = part). In the holological approach, an ecosystem is treated as a black box without scrutinizing the internal contents of the system, and the attention is focused on inputs and outputs to and from the ecosystem. On the other hand, in the merological approach, the components or parts of a system are studied in detail and then integrated into a whole system, if possible.

The pioneering work for the holological approach to lakes was done by Birge (1915) and Hutchinson (1957), and others extended this line of research. These studies are all from an energy viewpoint.

Holological and entropic studies of lake ecosystems are made by Aoki (1987b, 1989a, 1990a, 1995, 1998). As examples, two lakes, Lake Biwa and Lake Mendota, are considered, and the entropy productions of these lakes are compared. As shown in Section 7.2.2, the degree of eutrophication is related to entropy production, and the entropic nature of the time course of lake ecosystems is proposed, which may be generalized to biotic systems from individuals to ecosystems.

7.1 Methods of Calculation (Aoki 1987b, 1989a, 1998)

The outline of the methods of the calculation of entropy production in lakes is provided in this section. For the detailed physical basis of the methods, the reader is referred to Aoki (1982a, 1982b) and the literature cited there.

Energy flows between a lake and its surroundings are due to direct solar radiation E_{dr}, scattered (diffuse) solar radiation E_{sc}, reflected solar radiation E_{rf}, downward IR radiation $E_{i\downarrow}$, upward IR radiation $E_{i\uparrow}$, the evaporation of water E_{evp}, and heat conduction–convection E_{con}. The energy balance equation is written as the change in energy content of a lake ΔQ equal to the energy inflow minus energy outflow.

$$\Delta Q = (E_{dr} + E_{sc} - E_{rf}) + (E_{i\downarrow} - E_{i\uparrow}) + (-E_{evp} \pm E_{con}) \tag{7.1}$$

Each term of Eq. (7.1) has been measured for various lakes.

Entropy Principle for the Development of Complex Biotic Systems. DOI: 10.1016/B978-0-12-391493-4.00007-X

Associated with these energy flows are corresponding entropy flows, which are now discussed.

7.1.1 Direct Solar Radiation

The daily solar energy incident on a unit area of a horizontal surface just outside the Earth's atmosphere is given by (Gates, 1980; Sellers, 1965)

$$Q_e = \frac{86,400}{\pi} e_1 \left(\frac{\overline{d}}{d}\right)^2 (H - \tan H)\sin\phi \sin\delta \text{ J m}^{-2} \text{ day}^{-1} \tag{7.2}$$

where e_1 is the solar constant (e.g., Section 9.1; Liou, 1980)

$$e_1 = 1,353 \text{ J m}^{-2} \text{ s}^{-1} \tag{7.3}$$

where $\overline{d}$ and d are, respectively, the mean and instantaneous distances between the Earth and the Sun; H is the half-day length expressed in radians; ϕ is the latitude of the observation point; and δ is the solar declination.

Likewise, the daily solar entropy incident on a unit area of a horizontal surface just outside the Earth's atmosphere is given by

$$Q_s = \frac{86,400}{\pi} s_1 \left(\frac{\overline{d}}{d}\right)^2 (H - \tan H)\sin\phi \sin\delta \text{ J m}^{-2} \text{ day}^{-1} \text{ K}^{-1} \tag{7.4}$$

where s_1 is "the solar constant of second kind" (Aoki, 1983), which represents the solar entropy flux incident on a plane perpendicular to incident solar radiation at the top of the atmosphere. The value for s_1 is (Section 9.1; Aoki, 1983)

$$s_1 = 0.3132 \text{ J m}^{-2} \text{ s}^{-1} \text{ K}^{-1} \tag{7.5}$$

From Eqs (7.2) and (7.4),

$$Q_s = Q_e \times \frac{s_1}{e_1} \tag{7.6}$$

The energy of direct solar radiation incident on a horizontal plane of the Earth's surface is written as

$$E_{dr} = \rho Q_e \tag{7.7}$$

where ρ is the transmissivity of solar radiation in the atmosphere. Likewise, the entropy of direct solar radiation incident on a horizontal plane of the Earth's surface is

$$S_{dr} = \rho Q_s \tag{7.8}$$

From Eqs (7.6), (7.7), and (7.8),

$$S_{dr} = E_{dr} \times \frac{s_1}{e_1} = (2.31 \times 10^{-4}\ \mathrm{K}^{-1})E_{dr} \tag{7.9}$$

Thus, it is possible to calculate S_{dr} from the observed value for E_{dr} by use of the known values of e_1 and s_1. Equation (7.9) can be applied for any time interval, i.e., not only for the daily but also for the monthly or the annual direct solar entropy incident on the Earth's surface.

7.1.2 Diffuse Solar Radiation

Suppose that a beam of diffuse solar radiation is incident on an element of area $\mathrm{d}\sigma$ on a horizontal plane of the Earth's surface through a solid angle $\mathrm{d}\Omega = \sin\theta\, \mathrm{d}\theta\, \mathrm{d}\phi$ in a direction forming an angle θ with the normal to the area $\mathrm{d}\sigma$ (ϕ is an azimuthal angle of the incident radiation beam). Let the specific intensity (Planck, 1959, 1988) of the diffuse solar energy radiation be denoted as K_1. The radiation energy incident on $\mathrm{d}\sigma$ in time $\mathrm{d}t$ through $\mathrm{d}\Omega$ is, by definition, $K_1\, \mathrm{d}t\, \mathrm{d}\sigma \cos\theta\, \mathrm{d}\Omega$. Now assume, for simplicity, that diffuse solar radiation comes from all directions in the sky hemisphere with equal intensity and that hence K_1 is independent of (θ, ϕ). Integrating $K_1\, \mathrm{d}t\, \mathrm{d}\sigma \cos\theta\, \mathrm{d}\Omega$ by $\mathrm{d}\Omega$, the total radiation energy incident on $\mathrm{d}\sigma$ in time $\mathrm{d}t$ is obtained as

$$K_1 \mathrm{d}t\, \mathrm{d}\sigma \int_0^{2\pi} \mathrm{d}\phi \int_0^{\pi/2} \cos\theta \sin\theta\, \mathrm{d}\theta = \pi K_1\ \mathrm{d}t\ \mathrm{d}\sigma \tag{7.10}$$

Thus, the energy of diffuse (scattered) solar radiation incident per unit time per unit area of the Earth's surface is expressed as

$$E_{sc} = \pi K_1 \tag{7.11}$$

By means of Eq. (7.11), it is possible to calculate K_1 from the observed value for E_{sc}. Even if K_1 is dependent on (θ, ϕ) as will really be the case, Eq. (7.11) will remain valid by interpreting K_1 as representing a kind of average value.

Next, let the specific intensity of solar energy radiation in extraterrestrial space be K_0. Let us assume that extraterrestrial solar radiation is black-body radiation of the temperature $T_0 = 5{,}760$ K (Aoki, 1983). Then, by means of Planck's formula (Planck, 1959, 1988), K_0 is given by

$$K_0 = \int_0^{\infty} \frac{2h}{c^2} \frac{1}{e^{h\nu/kT_0} - 1} \nu^3 \mathrm{d}\nu = \frac{1}{\pi}\sigma T_0^4 = 0.63 \times 10^6\ \mathrm{G\,J\,m^{-2}\,year^{-1}} \tag{7.12}$$

where σ is the Stefan–Boltzmann constant.

Since extraterrestrial solar radiation with K_0 is scattered by particles in the atmosphere and then becomes diffuse solar radiation with K_1, it is possible to write K_1 as proportional to K_0, as in Aoki (1982a),

$$K_1 = \varepsilon K_0 = \varepsilon \frac{1}{\pi} \sigma T_0^4 \tag{7.13}$$

Thus, in this model, diffuse solar radiation can be regarded, as gray-body radiation emitted by the body of the temperature T_0 and of the emissivity ε (Aoki, 1982a). Therefore, the specific intensity of diffuse solar entropy radiation is obtained as (Aoki, 1982a)

$$L_1 = \frac{1}{\pi} \frac{4}{3} \varepsilon \sigma T_0^3 X(\varepsilon) \tag{7.14}$$

where $\varepsilon = (K_1/K_0)$ and

$$X(\varepsilon) = \frac{45}{4\pi^4} \frac{1}{\varepsilon} \int_0^\infty y^2[(x+1)\ln(x+1) - x \ln x]\, dy \tag{7.15}$$

$$x = \frac{\varepsilon}{e^y - 1} \tag{7.16}$$

From Eqs (7.11), (7.13), and (7.14),

$$L_1 = \frac{4}{3} \frac{K_1}{T_0} X(\varepsilon) = \frac{1}{\pi} \frac{4}{3} \frac{E_{sc}}{T_0} X(\varepsilon) \tag{7.17}$$

In an approach similar to that used with Eq. (7.11), the entropy of diffuse solar radiation incident per unit time on a unit area of the Earth's surface is given by

$$S_{sc} = \pi L_1 \tag{7.18}$$

From Eqs (7.17) and (7.18),

$$S_{sc} = \frac{4}{3} \frac{E_{sc}}{T_0} X(\varepsilon) \tag{7.19}$$

Since $\varepsilon = (K_1/K_0)$ is known from T_0 [Eq. (7.12)] and E_{sc} [Eq. (7.11)], the value S_{sc} can be found from the observed value for E_{sc} and from T_0.

7.1.3 Diffusely Reflected Solar Radiation

Suppose that solar radiation is reflected by an element of area $d\sigma$ on a horizontal plane of the Earth's surface and is emitted into a solid angle $d\Omega = \sin\theta \, d\theta \, d\phi$ in a direction forming an angle θ with the normal to the area $d\sigma$ (ϕ is an azimuthal angle of reflected radiation). Let the specific intensity (Planck, 1959, 1988) of the reflected solar energy radiation be denoted as K_1. The radiation energy reflected by $d\sigma$ in time dt into $d\Omega$ is $K_1\, dt\, d\sigma \cos\theta\, d\Omega$, by definition. It is assumed that reflection is diffuse and that K_1 is independent of (θ, ϕ).

Integrating $K_1\, dt\, d\sigma \cos\theta\, d\Omega$ by $d\Omega$, we obtain the total radiation energy reflected diffusely by $d\sigma$ in time dt:

$$K_1 dt\, d\sigma \int_0^{2\pi} d\phi \int_0^{\pi/2} \cos\theta \sin\theta d\theta = \pi K_1\, dt\, d\sigma \tag{7.20}$$

Thus, the energy of solar radiation reflected diffusely per unit time by a unit area of the Earth's surface is given by

$$E_{rf} = \pi K_1 \tag{7.21}$$

When K_1 is dependent on (θ, ϕ), the specific intensity K_1 in Eq. (7.21) should be interpreted as representing a kind of average value.

In a manner similar to that in Section 7.1.2, let us express K_1 as proportional to K_0 (Aoki, 1982a):

$$K_1 = \varepsilon K_0 = \varepsilon \frac{1}{\pi} \sigma T_0^4 \tag{7.22}$$

where K_0 is the specific intensity of extraterrestrial solar radiation, and $T_0 = 5{,}760$ K is the temperature of the Sun. Thus, diffusely reflected solar radiation can be regarded as gray-body radiation emitted by the body of the temperature T_0 and of the emissivity ε (Aoki, 1982a). Therefore, the specific intensity of diffusely reflected entropy radiation is obtained as (Aoki, 1982a)

$$L_1 = \frac{1}{\pi}\frac{4}{3} \varepsilon \sigma T_0^3 X(\varepsilon) \tag{7.23}$$

where $\varepsilon = (K_1/K_0)$ and $X(\varepsilon)$ is given by Eqs (7.15) and (7.16). From Eq. (7.21) to Eq. (7.23),

$$L_1 = \frac{4}{3}\frac{K_1}{T_0} X(\varepsilon) = \frac{1}{\pi}\frac{4}{3}\frac{E_{rf}}{T_0} X(\varepsilon) \tag{7.24}$$

As with Eq. (7.21), the entropy of solar radiation reflected diffusely per unit time by a unit area of the Earth's surface is given by

$$S_{rf} = \pi L_1 \tag{7.25}$$

From Eqs (7.24) and (7.25),

$$S_{rf} = \frac{4}{3}\frac{E_{rf}}{T_0}X(\varepsilon) \tag{7.26}$$

Since $\varepsilon = (K_1/K_0)$ is known from T_0 [Eq. (7.12)] and E_{rf} [Eq. (7.21)], the value of S_{rf} is found from the observed value for E_{rf} and from T_0.

7.1.4 IR Radiation Emitted by the Lake Surface

The lake surface can be regarded as an emitter of gray-body radiation of a temperature T_w and of an emissivity ε_w (Sellers, 1965). The entropy flux of gray-body radiation emitted by a gray body of the temperature T_w and the emissivity ε_w is written as [see Eq. (7.14) or (7.23), and

$$\text{the entropy flux} = \int_0^{2\pi} d\phi \int_0^{\pi/2} L_1 \cos\theta \sin\theta \, d\theta = \pi L_1]$$

$$S_{i\uparrow} = \frac{4}{3}\varepsilon_w \sigma T_w^3 X(\varepsilon_w) \tag{7.27}$$

The IR emissivity of the lake surface is taken as $\varepsilon_w = 0.94$, according to Sellers (1965). Thus, the entropy flux of IR radiation emitted by the lake surface can be calculated by using the observed value of the temperature of the lake surface T_w.

7.1.5 IR Radiation Incident on the Lake Surface

Let us assume that the atmosphere can be regarded as a gray body of an effective temperature T_a and an emissivity ε_a, as in Aoki (1988b). Then the energy flux of IR radiation emitted by the atmosphere is $E_{i\downarrow} = \varepsilon_a \sigma T_a^4$ (Aoki, 1988b). Since the atmosphere absorbs 94% of IR radiation emitted by the Earth's surface (Battan, 1979) and the absorptivity equals the emissivity (Kirchhoff's Law), the IR emissivity $\varepsilon_a = 0.94$. The effective temperature T_a can be obtained from ε_a and the observed IR radiation energy $E_{i\downarrow}$,

$$T_a = [E_{i\downarrow}/\varepsilon_a \sigma]^{1/4} \tag{7.28}$$

The entropy flux of IR radiation emitted by the atmosphere and incident on the lake surface is given by [as with Eq. (7.27)]

$$S_{i\downarrow} = \frac{4}{3}\varepsilon_a \sigma T_a^3 X(\varepsilon_a) = \frac{4}{3}\frac{E_{i\downarrow}}{T_a}X(\varepsilon_a) \tag{7.29}$$

the value of which is obtained from the known values $E_{i\downarrow}$, T_a, and ε_a.

7.1.6 Evaporation of Water and Heat Conduction–Convection

The entropy flux from the lake surface to the atmosphere due to the evaporation of water is

$$S_{evp} = \frac{E_{evp}}{T_w} \tag{7.30}$$

where E_{evp} is the latent heat flux from the lake surface, and T_w is the temperature of the lake surface.

Likewise, the entropy flux between the lake surface and the atmosphere due to heat conduction–convection is given by

$$S_{con} = \frac{E_{con}}{T_w} \tag{7.31}$$

where E_{con} is the heat flux due to conduction–convection.

Thus, the values of S_{evp} and S_{con} are obtained from observed data E_{evp}, E_{con}, and T_w.

7.1.7 Entropy Production

The net entropy inflow into the lake is obtained as

$$\Delta_e S = (S_{dr} + S_{sc} - S_{rf}) + (S_{i\downarrow} - S_{i\uparrow}) + (-S_{evp} \pm S_{con}) \tag{7.32}$$

The change in entropy content of the lake is (Aoki, 1989a, 1990a)

$$\Delta S = \frac{\text{the change in heat storage in the lake}}{T_m} \tag{7.33}$$

where T_m is the mean temperature of lake water. Thus, the value of ΔS is estimated from observed data of T_m and the change in heat storage in the lake water. The mean temperature is obtained by summing the product of the temperature at a series of depths of water layer and the volume fraction of that layer, and the change of heat storage in lake water is also estimated by observations (Aoki, 1989a, 1990a).

The entropy balance equation is expressed as

$$\Delta S = \Delta_e S + \Delta_i S \tag{7.34}$$

where $\Delta_i S$ is the entropy production within the lake. The entropy production is the production rate of entropy by the irreversibility of processes, and it is nonnegative according to the Second Law of Thermodynamics [Eq. (1.2)]. Because almost all processes occurring in the natural world are irreversible, the entropy production is a measure of the magnitude of activity of physical, chemical, and biological processes in the natural world, as stated in Chapter 1.

From Eq. (7.34), the entropy production can be obtained when values of the entropy flow ($\Delta_e S$) and the change in entropy content (ΔS) are estimated.

The methods of calculation shown here are based on simplified models in order to give clear, simple, and pertinent pictures on the subject. More complicated and detailed treatments may be possible but are left for future study. This is according to well-established methodologies in physics.

7.2 Lake Biwa and Lake Mendota

Consider two lakes as examples: Lake Biwa and Lake Mendota. Lake Biwa is located at 34°58′−35°31′N, 135°52′−136°17′E (near Kyoto, Japan) and consists of a northern basin (the main part) and a southern basin (the smaller part). The former is oligotrophic, and the latter is nearly eutrophic (in the 1970s; data of that period are used in the present study). Only the northern basin is considered here. Lake Biwa is the most studied lake in Japan, and its energy budgets (i.e., energy flow and heat storage) were studied in detail by Ito and Okamoto (1974) and Kotoda (1977).

From the data given by Kotoda (1977), monthly entropy productions per surface area unit in Lake Biwa are obtained using the methods shown in Section 7.1 similar to a previous article (Aoki, 1990a). The results, entropy production per square meter of lake surface plotted against absorbed solar radiation energy per square meter of lake surface, are shown in Figure 7.1.

Lake Mendota, another example, is located at 43°04′N, 89°24′W (near Madison, Wisconsin, USA) and is a eutrophic lake. It is the most studied lake in the USA, and its energy budgets were investigated by Dutton and Bryson (1962) and Stewart (1973). From their data, monthly entropy productions per surface area in Lake Mendota are obtained by similar methods employed in Section 7.1 and in the previous paper (Aoki, 1989a). The results are shown in Figure 7.2.

Figures 7.1 and 7.2 show that the entropy production in month j [denoted as $(\Delta_i S)_j$] is a linear function of the absorbed solar radiation energy in month j (denoted as E_j):

$$(\Delta_i S)_j = a + bE_j \tag{7.35}$$

The second term on the right-hand side of Eq. (7.35) is the entropy production dependent on solar radiation energy. This term is mainly due to the absorption of solar radiation by water, dissolved organic matter, and suspended solid (SS) in the lake water followed by subsequent conversion of solar radiation energy mostly to heat energy; this process causes entropy production. (Contributions to this term from photosynthesis and light respiration of phytoplankton will be very small compared with these mentioned.) The first term on the right-hand side of Eq. (7.35) is

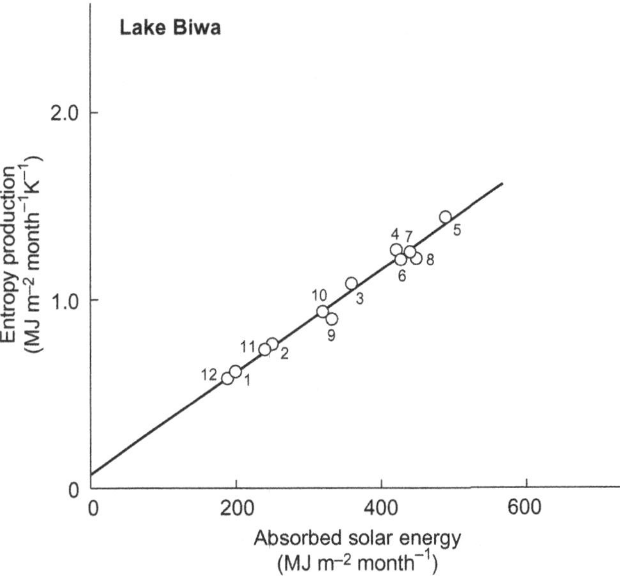

Figure 7.1 Monthly entropy production in the northern basin of Lake Biwa per square meter of the lake surface plotted against monthly solar radiation energy absorbed by 1 m^2 of the lake surface. The numbers near the circles are the months.

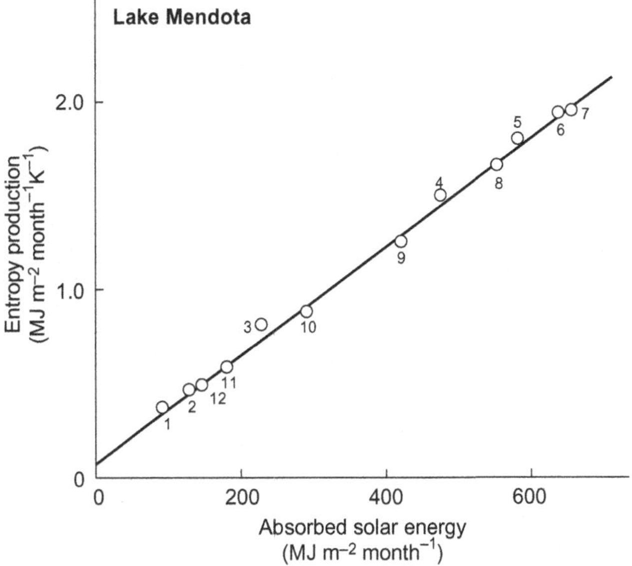

Figure 7.2 Monthly entropy production in Lake Mendota per square meter of the lake surface plotted against monthly solar radiation energy absorbed by 1 m^2 of the lake surface. The numbers near the circles are the months.

the entropy production independent of solar radiation energy. This term is assumed, for the moment, to be mainly due to the liberation of heat energy by the respiration of organisms in the lake; this process leads to entropy production.

The entropy production per year is given by

$$\sum_{j=1}^{12} (\Delta_i S)_j = 12a + b \sum_{j=1}^{12} E_j \tag{7.36}$$

The value of each term in Eqs (7.35) and (7.36) is easily obtained from the data in Figures 7.1 and 7.2.

Figure 7.3 shows the distributions of entropy production in Lake Biwa per year per square meter of lake surface, which are derived from the data in Figure 7.1. The absorbed solar radiation energy per year per square meter of the lake surface is 4,153 MJ. Consider a water column from the lake surface to the bottom, the cross section of which is 1 m^2. This water column consists of a light zone (euphotic zone) and a dark zone (aphotic zone). The entropy production dependent on solar radiation in the light zone is 11.1 MJ K^{-1} $year^{-1}$. The entropy production independent of solar radiation is 0.8 MJ K^{-1} $year^{-1}$, which is distributed over the whole water column because bacteria and other organisms are distributed even in the dark zone and benthic organisms in the bottom. A value in parentheses in Figure 7.3 is the corresponding entropy production per MJ of absorbed solar radiation energy in units of 10^{-3} K^{-1}:

$$11.1 \text{ MJ K}^{-1}/4,153 \text{ MJ} = 2.7 \times 10^{-3} \text{ K}^{-1} \tag{7.37}$$

Total entropy production per incident solar radiation energy per unit volume in the water column (see Figure 7.3) is

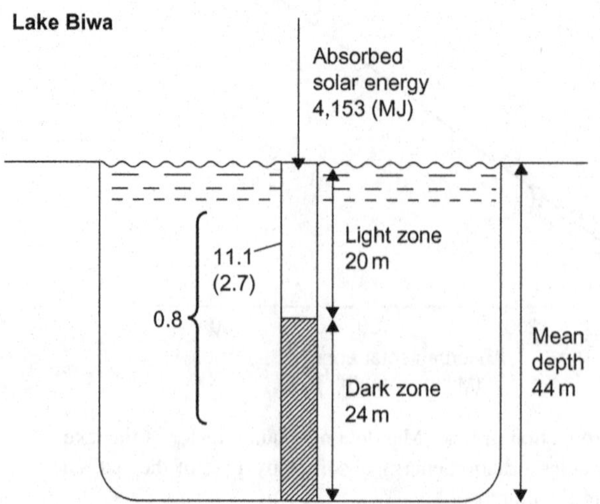

Figure 7.3 Pattern of entropy production per year in the northern basin of Lake Biwa. Explanations are given in the text.

$$11.9\ \mathrm{MJ\ K^{-1}\ m^{-2}}/4,153\ \mathrm{MJ}/44\ \mathrm{m} = 0.07\ \mathrm{m^{-3}\ K^{-1}} \tag{7.38}$$

Solar energy-dependent entropy production per incident solar radiation energy per unit volume of the light zone is

$$11.1\ \mathrm{MJ\ K^{-1}\ m^{-2}}/4,153\ \mathrm{MJ}/20\ \mathrm{m} = 0.13\ \mathrm{m^{-3}\ K^{-1}} \tag{7.39}$$

Solar energy independent entropy production per unit volume of the column is

$$0.8\ \mathrm{MJ\ K^{-1}\ m^{-2}}/44\ \mathrm{m} = 19\ \mathrm{kJ\ K^{-1}\ m^{-3}} \tag{7.40}$$

These values are shown in the first row in Table 7.1. Figure 7.4 shows the distribution of entropy produced in Lake Mendota, and values similar to those in

Table 7.1 Comparison of Entropy Productions in Lake Biwa and Lake Mendota

Lakes	Total (in Whole Water Column)	Solar Energy Dependent (in Light Zone)	Solar Energy Independent (in Whole Water Column)
Lake Biwa	0.07	0.13	19
Lake Mendota	0.24	0.31	69
Lake Mendota/Lake Biwa	3.7	2.3	3.6

Total and solar energy-dependent entropy productions (per year per MJ of absorbed solar radiation energy per cubic meter of the lake water) are shown, respectively, in the first and second columns; entropy productions independent of solar radiation energy (per year per cubic meter of the lake water) are in the third column in units of $\mathrm{kJ\ K^{-1}\ m^{-3}\ year^{-1}}$. Ratios of the values for the two lakes are shown in the last row. Explanations are given in the text.

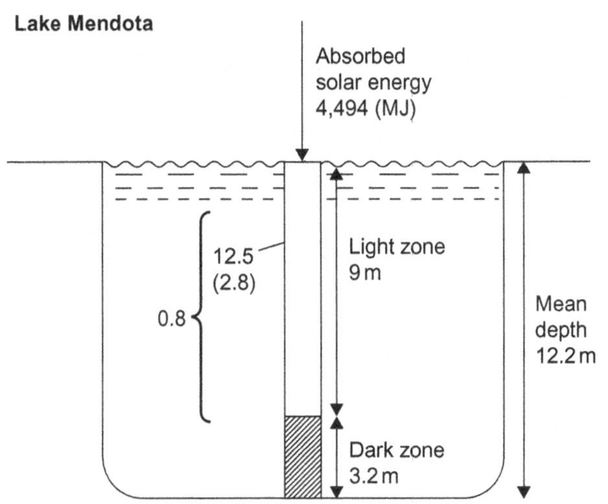

Figure 7.4 Pattern of entropy production per year in Lake Mendota, parallel to Figure 7.3.

Lake Biwa are shown in the second row in the table. The third row is of the ratios of the values in Lake Mendota/Lake Biwa.

7.2.1 Limnological Meanings

As already stated, the entropy production dependent on solar radiation in the light zone (euphotic zone) is the measure of activity due to the absorption of solar radiation energy by water, dissolved organic matter, and SS, followed by the subsequent conversion of solar energy to heat energy. Hence, the values of entropy production dependent on solar radiation in the light zone in Table 7.1 are related to the amount of dissolved organic matter and SS per cubic meter of lake water in the light zone. In fact, the average amount of SS in the light zone in Lake Biwa is 1.3 g m^{-3}; in Lake Mendota it is 1.9 g m^{-3} (National Institute for Research Advancement, 1984), and the ratio of the amount of SS in Lake Mendota to that in Lake Biwa is 1.5. Also, the average amount of dissolved organic carbon (DOC) in the light zone in Lake Biwa is 1.6 g C m^{-3} (Mitamura & Saijo, 1981); in Lake Mendota it is 3.3 g C m^{-3} (Brock, 1985); the ratio of DOC in Lake Mendota to that in Lake Biwa is 2.1. These ratios are consistent with the ratio of entropy production dependent on solar radiation in Lake Mendota to that in Lake Biwa, as shown in Table 7.1. Thus, the larger the amounts are of SS and DOC, the more the entropy production is dependent on solar radiation. Suspended solid is composed of planktons, fine broken pieces of dead organisms and their decomposed matter, excrements of organisms, bacteria attached to them, and abiotic (clay-like) small particles. It should be noted that the amount of SS depends on methods of measurement (e.g., pore-sized and/or microfilter quality). This is also the case for DOC. The entropy production dependent on solar radiation gives a kind of *physical* measure for the amount of dissolved organic matter and SS in the lake water by means of reactions to incident solar radiation. This measure is irrelevant to the difference in ordinary methods of measurement of SS and DOC.

The entropy production independent of solar radiation energy shown in the third column of Table 7.1 is the measure of activity of the respiration of organisms and is distributed over the whole water column from the surface to the bottom, as already stated. Hence, it correlates with the amount of organisms in the whole water column. In fact, the average amount of total plankton plus zoobenthos in the whole water column in Lake Biwa is 0.16 g C m^{-3} (Sakamoto, 1975); in Lake Mendota it is 0.62 g C m^{-3} (Brock, 1985); the ratio of the amount of plankton plus zoobenthos in Lake Mendota to that in Lake Biwa is 3.9. This ratio is consistent with the result on entropy production independent of solar radiation as shown in Table 7.1. The larger the amount of organisms is, the more the entropy production is independent of solar radiation. The entropy production independent of solar radiation energy represents a *physical* measure of the degree of respiration of organisms in lake water.

Thus, entropy productions due to light absorption (a physical process) and respiration (a biological process) are separately estimated and compared to each other for the oligotrophic and eutrophic lakes.

The values in the first column of Table 7.1 are total entropy productions per year per MJ of absorbed solar radiation energy per cubic meter of the average whole water column.

7.2.2 Increasing Entropy Production Principle

As shown in Table 7.1, the entropy productions in eutrophic Lake Mendota are larger than those in oligotrophic Lake Biwa in any of the categories considered (light absorption, respiration, and total). Therefore, it may be possible to propose that entropy production in a eutrophic lake will generally be larger than that in an oligotrophic lake. Eutrophication in a lake is a directional process: as stated in Chapter 6, the process tends to proceed with time from oligotrophy to eutrophy in most present lake ecosystems that are surrounded especially by the environment full of organic matter (anthropogenic restoration is not considered here). Hence, the entropy production in lakes will increase with time, accompanying the process of eutrophication; this may be called the Entropy Law for Eutrophication and has already been proposed by Aoki (1989a, 1990a). It should be noted that this principle of the increase of entropy production with time was pointed out by Aoki (1989a, 1990a) for the first time for real natural processes occurring in nature. However, "the increasing entropy production principle" is opposite to the Prigogine Minimum Entropy Production Principle (Nicolis & Prigogine, 1977), which states that entropy production decreases with time and reaches a minimum (Chapter 1). Prigogine's principle holds only near to the thermal equilibrium; however, it is known that ecosystems are far from equilibrium. Hence, it is neither surprising nor strange that the Prigogine's principle does not hold in ecological systems.

7.3 Senescent Stages

Will this principle, however, be applied to the overall span of ecological process? Hypereutrophic lakes represent the ultimate senescent stage of the eutrophication process; they become very vulnerable to external perturbations, and the entire algal biomass often catastrophically collapses and dies, resulting in the elimination of entire populations (phytoplankton, fish, and zooplankton) (Barica, 1993). Also, the productivity of organic matter in a senescent lake is lower than that in a eutrophic lake, as already shown in the classical work by Lindeman (1942). These facts suggest that entropy production will decrease with time in the senescent stage of the ecological process (i.e., in the course of hypereutrophication) because entropy production is a measure of the degree of various activities in natural systems, including productivity in ecological systems, as stated in Chapter 1.

Thus, the entropy production in ecological systems will have the tendency to initially increase and later decrease. This is in accord with the multi(three-)stage hypothesis proposed by Aoki (1989a) for the life span of the ecological process in general: entropy production increases with time in an initial stage (growing stage) of the ecological process (the entropy law for eutrophication holds in this stage), is kept almost stationary in an intermediate stage, and decreases in a later stage (senescent stage) of the ecological process. Thus, the ecological process will consist of multiphases in entropy production and not be unidirectional with the progress of ecosystems development.

7.4 Discussion

Ecosystems are complex and have various characteristic aspects, and there are many ecological theories from various viewpoints. Despite coming from various angles, these ecosystem theories are consistent and not in contradiction, as shown by the extensive study of Jørgensen (2002). This chapter deals with a rather narrow aspect, that is, an entropic view of how states of ecosystems change with time. This study is the first entropic study of ecosystems; it is not in contradiction with other ecosystem theories but rather is supplementary to them.

8 Entropy Principle in Living Systems (Min–Max Principle)

The multiphase (two- or three-phase) tendency in entropy production in the life span of plant leaves (Chapter 2), in pigs (Chapter 3), in humans (Chapter 5), and in ecological systems (Chapters 6 and 7) leads to an *entropy principle* in biotic systems from organisms to ecosystems. That is, entropy production in a biotic system in general, which is open and far from equilibrium, consists of two or more phases: an initial increasing stage, a later decreasing stage, and an intermediate stage, as shown schematically with possible fluctuations in Figure 8.1. In Figure 8.1 the initial point is the beginning of life in living systems (e.g., fertilization in organisms or the formation of the lake in lake ecosystems), and the last point is the end of their lives (death).[1] The initial increasing stage is the systems' phase of development and growth; the later decreasing stage is the phase of senescence and decay of the systems; and the intermediate stage is the transitional one (it may be stationary or oscillating). This entropy principle, the so-called Min–Max Principle (Aoki, 2006, 2008), will be universal for the development of living systems from organisms to ecosystems, which have two opposing phases: birth–growth and senescence–death (Aoki, 1989a, 1995, 1998).

The three-stage character of the entropy principle, shown in Figure 8.1 without fluctuations, is in accord with the classical hypothesis by Lindeman (1942) on the productivity of organic matter in lake ecosystems. Lindeman presented a similar hypothetical figure on the time-course of productivity, based on the comparison of senescent Ceder Bog Lake and eutrophic Lake Mendota. Entropy production is a measure of the degree of physical, chemical, and biological activities in natural systems (including the productivity in ecosystems), as stated in Chapter 1. Consequently, the entropy principle can be regarded as a kind of generalization and thermodynamical abstraction of Lindeman's classical hypothesis on the productivity of organic matter in lake ecosystems.

Of course, time scale may be a geological one in ecosystems, as in communities (Chapter 6).

The Min–Max Principle of Entropy Production (MMEP) is a gross and macroscopic pattern of the overall trend of the life span from the beginning to the end of living systems (even ecosystems have an end of life after a long period of time). On a small scale and over a short period of time, entropy production may show fluctuations and various patterns. They may only increase or only decrease, or they

[1] After the death of lakes, new ecosystems of wetlands, grasslands, etc. appear.

Entropy Principle for the Development of Complex Biotic Systems. DOI: 10.1016/B978-0-12-391493-4.00008-1

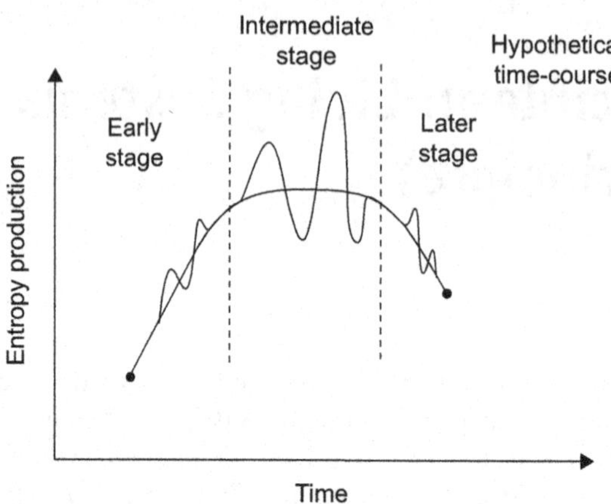

Figure 8.1 A hypothetical entropy principle for biotic systems. The scales are arbitrary. Explanations are given in the text.

may fall into irregular patterns (e.g., exercise and chills in humans, Chapter 4, oscillations in plant leaves with a daily arch, Chapter 2, or oscillations in lake ecosystems with an annual arch, Chapter 7). In the time scale of an hour, entropy productions in humans and lizards are high in active periods and low in rest periods (Section 4.4; Lamprecht, 2003 and the literature cited therein). The study of the fine structure of entropy production in a short period of time is of much significance. But they are short-term patterns and cannot be extended to the overall behavior. When fluctuations are accidentally large and entropy production becomes zero, systems tend to be destroyed.

Thus, various types of MMEPs, with fluctuations, are essential features of the life span of most living systems.

Many series of continuations, interactions, and overlapping of the structures of the various types of entropy production, with internal fluctuations and variations in the external effects, are the driving forces for the evolution of living systems. Thus the Min–Max Principle, with fluctuations, is the prototype for biological and ecological evolution, though it is impossible to fully develop the arguments here.

9 The Earth

The Earth's surface and atmosphere make up the biggest ecosystem, but it is the only one (at present) and a comparative study is impossible. However, entropy production on the Earth can be calculated. The calculation is based on a simple physical model and on observed energy budgets on the Earth. Entropy productions on this and the other planets of the solar system are discussed in Section 9.1 (Aoki, 1983) without the effects of beam splitting (Aoki, 1982b), diffuse reflection, and the scattering of solar radiation (Aoki, 1982a). Results on entropy flows and entropy productions on the Earth's surface and in the atmosphere, including effects of beam splitting, diffuse reflection, and scattering (Aoki, 1982a, 1982b), are treated in a simple way, giving only the results in Section 9.2. For full detailed arguments, the reader is referred to Aoki (1988b).

9.1 Planets of the Solar System

9.1.1 Solar Entropy Flow

The Sun can be regarded as a black-body sphere of a temperature T and of a radius r. The radiation energy emitted per unit time from the unit area of the surface of the Sun is, according to the Stefan–Boltzmann Law, σT^4, and the corresponding radiation entropy emitted is $(4/3)\sigma T^3$ (Planck, 1959, 1988), where σ is the Stefan–Boltzmann constant. The total energy E emitted from the Sun per unit time is given by $E = 4\pi r^2 \sigma T^4$, and the total entropy S by $S = 4\pi r^2\ (4/3)\ \sigma T^3 = (4/3)E/T$. If there are no energy losses in space, the quantity E equals the energy E_1 that passes in unit time through a larger sphere of a radius $R(>r)$. Likewise, if radiation is not absorbed, reflected, or scattered in space, the entropy of radiation does not increase in space, and the quantity S equals the entropy S_1 that passes in unit time through the sphere. Let $E_1 = 4\pi R^2 e_1$ and $S_1 = 4\pi R^2 s_1$, where e_1 is the energy flux through the sphere and s_1 the corresponding entropy flux. From $E = E_1$ and $S = S_1$, the energy flux e_1 and the entropy flux s_1, are expressed as follows:

$$e_1 = \left(\frac{r}{R}\right)^2 \sigma T^4 \tag{9.1}$$

$$s_1 = \left(\frac{r}{R}\right)^2 \frac{4}{3} \sigma T^3 = \frac{4}{3}\frac{e_1}{T} \tag{9.2}$$

Let us consider R as the mean distance between the Sun and the Earth. Then e_1 is the solar energy flux at the top of the atmosphere, and s_1 is the corresponding

Entropy Principle for the Development of Complex Biotic Systems. DOI: 10.1016/B978-0-12-391493-4.00009-3

solar entropy flux. The solar energy flux e_1 (the so-called solar constant) has been measured and given as $e_1 = 1.94\ \text{cal cm}^{-2}\ \text{min}^{-1} = 0.1353\ \text{J cm}^{-2}\ \text{s}^{-1}$ (Liou, 1980). Then, from Eqs (9.1) and (9.2), the effective temperature of the Sun, T, and the solar entropy flux, s_1, are obtained:

$$T = \left[\left(\frac{R}{r}\right)^2 \frac{e_1}{\sigma}\right]^{1/4} = 5,760\ \text{K} \tag{9.3}$$

$$s_1 = \frac{4}{3}\frac{e_1}{T} = 3.132 \times 10^{-5}\ \text{J cm}^{-2}\ \text{s}^{-1}\ \text{K}^{-1} \tag{9.4}$$

The solar entropy flux s_1 may be called the "solar constant of second kind," corresponding to e_1, the solar constant. If solar energy and entropy are spread uniformly over the full surface of the Earth, the energy received per unit area per unit time at the top of the atmosphere is given by

$$\begin{aligned}\bar{e}_1 &= \frac{\pi r_1^2 e_1}{4\pi r_1^2} = \frac{1}{4}e_1 \\ &= 3.383 \times 10^{-2}\ \text{J cm}^{-2}\ \text{s}^{-1} \\ &= 1,067\ \text{k J cm}^{-2}\ \text{year}^{-1}\end{aligned} \tag{9.5}$$

and the corresponding entropy by

$$\begin{aligned}\bar{s}_1 &= \frac{\pi r_1^2 s_1}{4\pi r_1^2} = \frac{1}{4}s_1 \\ &= 7.830 \times 10^{-6}\ \text{J cm}^{-2}\ \text{s}^{-1}\ \text{K}^{-1} \\ &= 246.9\ \text{J cm}^{-2}\ \text{year}^{-1}\ \text{K}^{-1}\end{aligned} \tag{9.6}$$

where r_1 is the radius of the Earth.

9.1.2 Entropies on the Planets of the Solar System

Let us consider the Earth and its atmosphere as a whole and regard it for IR radiation as a black-body sphere of a temperature T_1 and of a radius r_1. The energy and the entropy emitted per unit time from the unit area of the surface of the Earth planet are $e_2 = \sigma T_1^4$ and $s_2 = (4/3)\sigma T_1^3 = (4/3)e_2/T_1$, respectively. If it is assumed that the Earth's surface and atmosphere are in a steady state in energy, as they should be from a short-term point of view, absorbed energy should balance with emitted energy. The balance equation for the absorption and emission of radiation energy is

$$(1-a)\pi r_1^2 e_1 = 4\pi r_1^2 e_2 \tag{9.7}$$

that is,

$$e_2 = \frac{1}{4}(1 - a)e_1 \tag{9.8}$$

where a is the albedo, that is, the ratio of reflected to incident solar radiation for the planet. Since $a \simeq 0.30$ from observations (Sellers, 1965), the energy flux e_2 is given as $e_2 = 2.368 \times 10^{-2}$ J cm^{-2} s^{-1}. Then the effective temperature of the planet T_1 and the entropy flux s_2 are obtained:

$$T_1 = \left[\frac{e_2}{\sigma}\right]^{1/4} = 254 \text{ K} \tag{9.9}$$

$$s_2 = \frac{4}{3}\frac{e_2}{T_1} = 1.242 \times 10^{-4} \text{ J cm}^{-2} \text{ s}^{-1} \text{ K}^{-1} \tag{9.10}$$

From the values for s_1 and s_2 obtained in Eqs (9.4) and (9.10), the net amount of radiation entropy absorbed by the Earth per unit time is

$$(1 - a)\pi r_1^2 s_1 - 4\pi r_1^2 s_2 = -6.055 \times 10^{14} \text{ J s}^{-1} \text{ K}^{-1} \tag{9.11}$$

If it is spread uniformly over the full surface of the Earth, the amount received per unit area is given as

$$\begin{aligned}\frac{\mathrm{d_e}S}{\mathrm{d}t} &= \frac{1}{4\pi r_1{}^2}\left[(1 - a)\pi r_1^2 s_1 - 4\pi r_1^2 s_2\right] \\ &= (1 - a)\bar{s}_1 - s_2 \\ &= -1.187 \times 10^{-4} \text{J cm}^{-2} \text{ s}^{-1} \text{ K}^{-1} \\ &= -3{,}744 \text{ J cm}^{-2} \text{ year}^{-1} \text{ K}^{-1}\end{aligned} \tag{9.12}$$

Thus, the flow of radiation entropy going out from the Earth is larger than that flowing into it, and the net flow of radiation entropy into the Earth is negative: the Earth, like biological organisms, absorbs "negative entropy" from its surroundings. It follows that the Earth has some possibility to develop organized structures on the surface and in the atmosphere (Schrödinger, 1944). The negative value of the entropy flow in Eq. (9.12) is the basis, from the entropy point of view, for the existence of organized structures on the Earth.

If it is assumed that the Earth surface and its atmosphere are in a steady state in entropy, then an entropy production $\mathrm{d_i}S/\mathrm{d}t$ should occur so as to add to the entropy flow [Eq. (9.12)] to result in no change of total entropy:

$$\begin{aligned}\frac{\mathrm{d_i}S}{\mathrm{d}t} &= -\frac{\mathrm{d_e}S}{\mathrm{d}t} = 1.187 \times 10^{-4} \text{ J cm}^{-2} \text{ s}^{-1} \text{ K}^{-1} \\ &= 3{,}744 \text{ J cm}^{-2} \text{ year}^{-1} \text{ K}^{-1}\end{aligned} \tag{9.13}$$

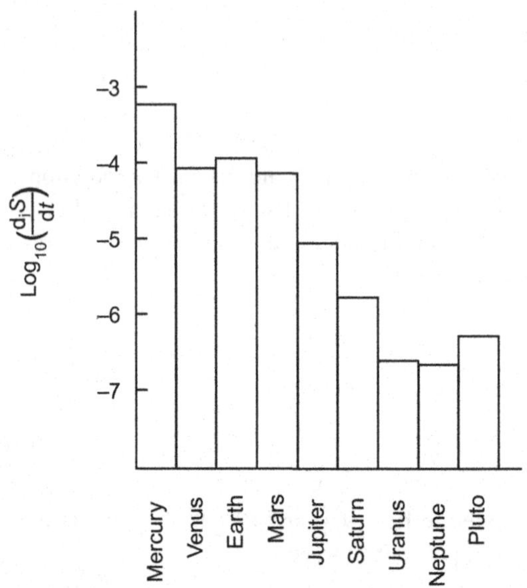

Figure 9.1 Log_{10} of the entropy production d_iS/dt for the planets of the solar system.

Similar calculations are carried out for other planets of the solar system. The results are shown in Figure 9.1. Data for the calculations of the other planets are in table I of Aoki (1983). The orders of magnitude of entropy productions are, in units of $\text{J cm}^{-2}\,\text{s}^{-1}\,\text{K}^{-1}$, 10^{-3} for Mercury, 10^{-4} for Venus, Earth, and Mars, and 10^{-5}–10^{-7} for the other outer planets. Comparative study suggests that the occurrence of the entropy production of the order of magnitude 10^{-4} will be one of the *necessary* conditions for the existence of the organized structures that exist on the Earth surface and in its atmosphere.

The present section is intended to provide, without going into the details of physics, clear and pertinent pictures. This treatment is based on a simplified model, neglecting entropy changes due to diffuse reflection and beam splitting of radiation (Aoki, 1982a, 1982b). It is possible to take into account these points and refine the calculations on the basis of a more sophisticated physical theory, which is treated in Aoki (1988b).

9.2 Earth's Surface and Atmosphere

A more detailed picture of entropy production on the Earth can be obtained by using the energy-flow diagram of the Earth's surface and atmosphere (Figure 9.2) and including the effects of beam splitting, diffuse reflection, and the scattering of radiation. The energy-flow diagram of the Earth takes the form of a two-compartment model (the surface and the atmosphere). The units are $\text{kJ cm}^{-2}\,\text{year}^{-1}$. The figures are taken from Sellers (1965), U.S. Committee for the Global Atmospheric Research Program (1975), and Battan (1979).

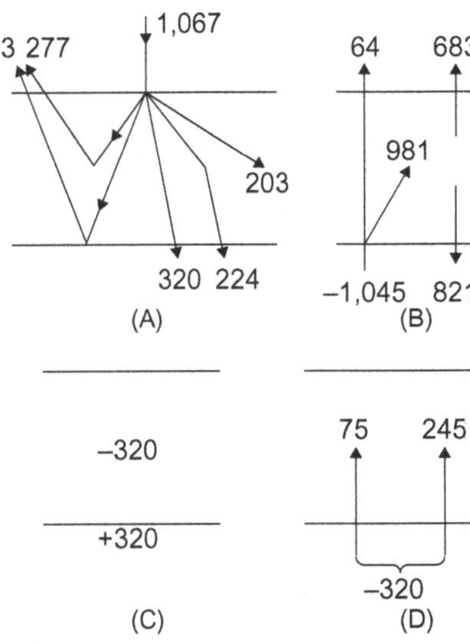

Figure 9.2 Energy flows (arrows) on the Earth. Figures are in units of kJ cm^{-2} $year^{-1}$. Explanations are given in the text.

- *Energy flows associated with incident solar radiation.* Of the incident energy flow 1,067 [Eq. (9.5)],
  - 320 (30% of the incident) are absorbed by the Earth's surface as direct-beam solar radiation,
  - 224 (21%) are absorbed by the Earth's surface as diffuse sky radiation,
  - 203 (19%) by the atmosphere,
  - Of the remainder, 277 (26%) are reflected back by the atmosphere, and 43 (4%) are reflected by the Earth's surface.
- *Energy flows associated with terrestrial IR radiation.* The Earth's surface emits 1,045, of which 64 (6.1% of those emitted) escape to outer space through so-called the atmospheric window. The remaining 981 (93.9%) are absorbed by the atmosphere. The atmosphere emits 683 to outer space and 821 to the Earth's surface.
- *Net radiation of the Earth's surface and atmosphere.* This is the net sum of the values in Figure 9.2A and B.
- *Energy flows due to heat conduction and due to evaporation of water.* Heat conduction transports 75 of sensible heat, and evaporation removes 245 of latent heat from the Earth's surface to the atmosphere.

Based on these energetic data, detailed patterns for entropy flow and entropy production in the Earth's surface and atmosphere are calculated by means of some physical procedures (Aoki, 1988b). The details are not described here due to the many physical concepts and equations; the interested reader is referred to the article by Aoki (1988b). The results of the calculation are shown in Figure 9.3, in units of J cm^{-2} $year^{-1}$ K^{-1}. As shown, entropy production in the Earth's surface is 2,064 and that in the atmosphere is 2,236.

The entropy flow 4,348 from the atmosphere to the Earth's surface consists of the entropy flow due to IR radiation (4,225) and the flow due to diffuse sky radiation (123). The entropy flow 582 from the Earth's surface to outer space consists of the entropy flow due to IR radiation (532) and that due to reflected solar radiation (50). The entropy flow 5,945 from the Earth's surface to the atmosphere consists of the

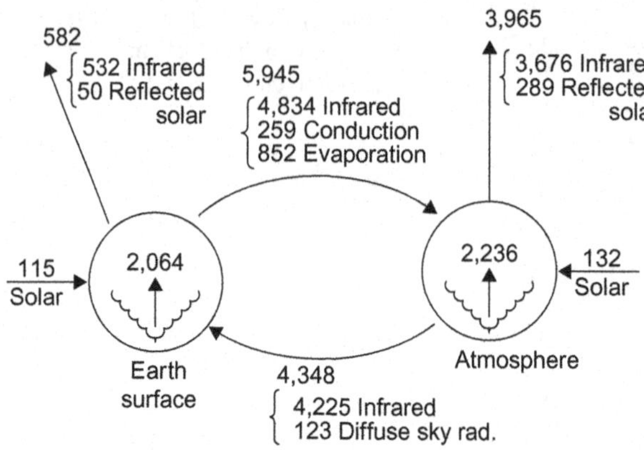

Figure 9.3 Entropy flows and entropy productions on the Earth's surface and in the atmosphere. Units are J cm^{-2} $year^{-1}$ K^{-1}. Explanations are given in the text.

entropy flow due to IR radiation (4,834), due to heat conduction (259), and due to evaporation of water (852). The entropy flow from the atmosphere to outer space (3,965) consists of the entropy flow due to IR radiation (3,676) and the flow due to reflected solar radiation (289). The total entropy flow from the Earth to outer space (4,547) becomes equal to the sum of the total entropy production on the Earth (4,300) and the incident solar entropy flow into the Earth (247) [Eq. (9.6)]. The net entropy flows from the Earth's surface to the atmosphere that are transported by IR radiation, heat conduction, and the evaporation of water are in the ratios of 1.0:0.4:1.4. There is an extensive circulation of entropy between the Earth's surface and atmosphere. The ratio of the entropy of incoming radiation to the Earth to that of outgoing radiation is 1:18; the Earth amplifies incoming radiation entropy by 18 times.

For the sake of simplicity and clarity, this section is based on the simplest pattern (the textbook level) of energy flows on the Earth (Figure 9.2). However, more detailed discussions on entropy on the Earth's surface and in its atmosphere may be developed if more detailed patterns of energy flow on the Earth are employed as the next step for entropy calculation.

Remark 1 Global Warming

It has been proposed, with a high probability, that human-made emissions of greenhouse gases induce global warming, leading to the change of climate (Collins, Colman, Haywood, Manning, & Mote, 2007; IPCC Report of WGI, 2007,), although some scientists oppose this view. Nevertheless, it can be said that effects of the change of climate may influence the pattern of energy flow on the Earth (Figure 9.2). Therefore, going forward, entropy production on the Earth may change; the energy flow may be different from Figure 9.2, in perhaps 20–50 years or more. At that time, it will be clear that entropy production has a tendency to increase or decrease. The entropy production on the Earth is a measure of its activity (Chapter 1), and so it would be desirable that entropy production is not in the decreasing phase. But we do not know that right now.

Remark 2 Gaia Hypothesis

According to the Gaia Hypothesis (Lovelock, 1979, 1988), the Earth is a self-controlled, harmonic, and interactive complex of biological and nonbiological worlds on the planet. It can be regarded metaphorically as if it were a kind of superorganism, and so it has been named Gaia. The entropy production (i.e., activities, Chapterl) of Gaia are given in Figure 9.3 as the sum of entropy productions on the Earth's surface and in the atmosphere. Its value is 4,300 J cm^{-2} $year^{-1}$ K^{-1}. Thus, the entropy production of Gaia is given by this concrete numerical value.

Gaia is the largest of ecosystems. As such, is it in an increasing zone of entropy production or a decreasing zone (Figure 8.1)? This is a question whose answer can have grave consequences for the future existence of all living systems on the Earth, including, of course, humans. However, no one knows the answer; it is an open question and left for future studies.

Appendix
Energy Budget Equation: The First Law of Thermodynamics

Wiegert (1968) developed thermodynamical considerations on the energy budgets of animals and claimed that energy budgets had a sound thermodynamic basis. Phillipson (1975) and Klekowski and Duncan (1975) also cited Wiegert's conclusions as the thermodynamical justification for the energy budget equation. However, the argument given by Wiegert (1968) is not complete and in reality does not constitute a proof of the equation for the energy budgets of animals: there is a gap of logic in his discussion because he did not derive the equation exactly from the First Law of Thermodynamics for open systems.

The concept of an energy budget is of importance in biology (Humphreys, 1979; International Biological Programme Publications Committee, 1967–1975), as is a precise understanding of the foundation on which it is based. In this appendix, an explicit derivation of the generalized equation is presented for energy budgets from the First Law of Thermodynamics, thus giving a solid thermodynamical basis to the equation for the energy budgets of living organisms. Thus, the argument put forward by Wiegert (1968) has been corrected and generalized here (Aoki, 1988c).

A.1 Fundamental Equation of Energy Budget

Let us start with the equation of the First Law of Thermodynamics for open systems (Beattie & Oppenheim, 1979; Haase, 1969):

$$dU = \delta Q + \delta W + \sum_{k} H_k d_e n_k \tag{A.1}$$

where dU is the increase in internal energy of the system, δQ is the heat absorbed by the system, δW is the work done on the system, H_k is the partial molar enthalpy of substance k, and $d_e n_k$ is the increase in the amount (in moles) of substance k due to the exchange of matter with its surroundings. Equation (A.1) also gives a definition of the "heat" δQ absorbed by an open system (Haase, 1969). The work δW done on the system consists of two parts, expansion work $\delta W_{EX} = -p dV$ and nonexpansion work δW_X:

$$\delta W = \delta W_{EX} + \delta W_X = -p\, dV + \delta W_X \tag{A.2}$$

where p is the pressure, and dV is the increase in volume of the system.

Let us consider biological organisms. For the treatment of the energy balance of organisms, it is convenient to consider only the energy change that is not due to expansion work. Expansion work is not useful for living organisms because it does not contribute directly to the maintenance of life (Wiegert, 1968). Therefore, it is better to adopt a more suitable quantity as an energy variable, that is, enthalpy $H = U + pV$, instead of the internal energy U of the system. The enthalpy change in the system, according to Eqs (A.1) and (A.2), is given by

$$\mathrm{d}H = \delta Q + V\,\mathrm{d}p + \delta W_{\mathrm{X}} + \sum_{\mathrm{k}} H_{\mathrm{k}}\mathrm{d}_{\mathrm{e}}n_{\mathrm{k}} \tag{A.3}$$

Since organisms for the most part are living under constant pressure ($\mathrm{d}p = 0$), Eq. (A.3) becomes

$$\mathrm{d}H = \delta Q + \delta W_{\mathrm{X}} + \sum_{\mathrm{k}} H_{\mathrm{k}}\mathrm{d}_{\mathrm{e}}n_{\mathrm{k}} \tag{A.4}$$

Generally, the heat absorbed by an organism δQ is expressed as

$$\delta Q = S + L_{\mathrm{in}} - L_{\mathrm{out}} \pm C \pm V \tag{A.5}$$

where S is the heat energy given to the organism by the absorption of visible light (e.g., solar radiation), L_{in} is the heat energy given to the organism by the absorption of infrared radiation, L_{out} is the heat energy of infrared radiation emitted from the organism to its surroundings, C is the heat energy emitted by conduction–convection [a plus (+) sign is adopted when the temperature of the organism is lower than that of its surroundings], and V is the heat energy emitted due to the evaporation of water [a plus (+) sign is adopted in the case of dew and frost formation on the surfaces of organisms (Nobel, 1970)]. The enthalpy change due to the exchange of matter with the surroundings $\sum_{\mathrm{k}} H_{\mathrm{k}}\mathrm{d}_{\mathrm{e}}n_{\mathrm{k}}$ equals the enthalpy of ingested matter I minus that of egested and excreted matter E:

$$\sum_{\mathrm{k}} H_{\mathrm{k}}\mathrm{d}_{\mathrm{e}}n_{\mathrm{k}} = I - E \tag{A.6}$$

The overall change in enthalpy $\mathrm{d}H$ is written as

$$\mathrm{d}H = P + K \tag{A.7}$$

where P is the chemical energy equivalent of net production (growth and reproduction), and K is the heat storage (positive or negative) in the body of the organism. Thus, Eq. (A.4) is expressed as

$$P + K = S + L_{\mathrm{in}} - L_{\mathrm{out}} \pm C \pm V + \delta W_{\mathrm{X}} + I - E \tag{A.8}$$

Table A.1 Examples of the Energy Budget Equation

Organism	Situations	Energy Budget	Reference
Plant leaves	In the daytime	$0 = S + L_{in} - L_{out} - C - V$	Nobel (1970)
Plant leaves	At night with frost formation	$0 = L_{in} - L_{out} + C + V$	Nobel (1970)
Plant leaves	At night	$K = L_{in} - L_{out} + C$	Schwintzer (1971)
Deer	On a winter night	$0 = L_{in} - L_{out} - C - V + I$	Moen (1968a, 1968b)
Lizard	In the daytime	$0 = S + L_{in} - L_{out} - C - V + I$	Bartlett and Gates (1967)
Human	In a respiration calorimeter	$K = L_{in} - L_{out} - C - V + I$	Hardy and DuBois (1938a)

S = the energy given to the organism by the absorption of visible light.
L_{in} = the energy given to the organism by the absorption of infrared radiation.
L_{out} = the energy of infrared radiation emitted by the organism.
C = the energy emitted by conduction–convection.
V = the energy emitted by evaporation of water.
K = the heat storage in the organism.
I = the metabolic heat produced by ingested food.

This is a generalized equation for the energy budget of living organisms derived from the First Law of Thermodynamics for open systems. It can be applied to any organism, irrespective of whether it is a plant or an animal. Some examples of the energy budget equation studied in the literature are shown in Table A.1.

A.2 Special Cases

When the temperature of the organism is higher than that of its environment, L_{in} is smaller than L_{out}, and $(L_{in} - L_{out} - C - V)$ is negative. Generally, the quantity $(L_{in} - L_{out} - C - V)$ becomes negative (although it is possible in some cases for this term to become zero or positive; Schwintzer, 1971). When it is negative, let $-(L_{in} - L_{out} - C - V)$ be denoted as R. When the organism is placed in almost complete darkness, as is often the case in measurements using a calorimeter, the quantity S becomes zero. Also, the nonexpansion work δW_X is negligible for most organisms except domestic draught animals (Wiegert, 1968). The quantity K is zero when the temperature of the organism is kept constant and does not change with time (K = heat capacity × temperature change). Under these conditions, Eq. (A.8) becomes

$$P = I - E - R \tag{A.9}$$

which is the usual form (Wiegert, 1968) of the equation for the energy budgets of animals.

This derivation makes it clear that the energy budgets of living organisms are clearly founded on the basis of the First Law of Thermodynamics for open systems. Thus, the argument given by Wiegert (1968) has been corrected and generalized. The conditions under which the usual expression [Eq. (A.9)] holds were also stated.

Obviously, the generalized energy budget equation [Eq. (A.8)] can be applied not only to organisms but also to populations, trophic levels, and whole ecosystems. For example, the energy budget for the northern basin of Lake Biwa is $K = S + L_{in} - L_{out} - C - V$ (Ito & Okamoto, 1974). Also, many studies based on the energy budget equation have been made in micrometeorology and in the study of global climate (Budyko, 1974; Monteith, 1973; and the literature cited therein).

It is clear from this derivation that the energy budget is an immediate consequence of the First Law of Thermodynamics but that it has no relation to the Second Law.

References

Allison, E. H., Patterson, G., Irvine, K., Thompson, A. B., & Menz, A. (1995). The pelagic ecosystem. In A. Menz (Ed.), *The fishery potential and productivity of the pelagic zone of Lake Malawi/Niassa* (pp. 351–367). Chatham: National Resources Institute.

Anderson, P. W. (1972). More is different. *Science*, *177*, 393–396.

Angelini, R., & Petrere, M., Jr. (1996). The ecosystem of Broa Reservoir, Sao Paulo State, Brazil, as described using ECOPATH. *Naga ICLARM Quarterly*, *19*, 36–41.

Aoki, I. (1982a). Radiation entropies in diffuse reflection and scattering and application to solar radiation. *Journal of the Physical Society of Japan*, *51*, 4003–4010.

Aoki, I. (1982b). Radiation entropies in beam-splitting and application to solar radiation. *Journal of the Physical Society of Japan*, *51*, 4011–4014.

Aoki, I. (1983). Entropy productions on the earth and other planets of the solar system. *Journal of the Physical Society of Japan*, *52*, 1075–1078.

Aoki, I. (1985) Systems analysis of animals from entropy point of view. Presentation at *International Symposium on Mathematical Biology*, Kyoto, Japan.

Aoki, I. (1987a). Entropy budgets of deciduous plant leaves and a theorem of oscillating entropy production. *Bulletin of Mathematical Biology*, *49*, 449–460.

Aoki, I. (1987b). Entropy balance in Lake Biwa. *Ecological Modelling*, *37*, 235–248.

Aoki, I. (1987c). Entropy budgets of soybean and bur oak leaves at night. *Physiologia Plantarum*, *71*, 293–295.

Aoki, I. (1987d). Entropy balance of white-tailed deer during a winter night. *Bulletin of Mathematical Biology*, *49*, 321–327.

Aoki, I. (1988a). Eco-physiology of a lizard (*Sceloporus occidentalis*) from an entropy viewpoint. *Physiology and Ecology of Japan*, *25*, 27–38.

Aoki, I. (1988b). Entropy flows and entropy productions in the earth's surface and in the earth's atmosphere. *Journal of the Physical Society of Japan*, *57*, 3262–3269.

Aoki, I. (1988c). Exact derivation of an energy budget equation on the basis of the First Law of Thermodynamics. *Ecological Research*, *3*, 53–56.

Aoki, I. (1989a). Holological study of lakes from an entropy viewpoint—Lake Mendota. *Ecological Modelling*, *45*, 81–93.

Aoki, I. (1989b). Entropy budget of conifer branches. *Botanical Magazine Tokyo*, *102*, 133–141.

Aoki, I. (1989c). Entropy flow and entropy production in the human body in basal conditions. *Journal of Theoretical Biology*, *141*, 11–21.

Aoki, I. (1990a). Monthly variations of entropy production in Lake Biwa. *Ecological Modelling*, *51*, 227–232.

Aoki, I. (1990b). Effects of exercise and chills on entropy production in human body. *Journal of Theoretical Biology*, *145*, 421–428.

Aoki, I. (1991). Entropy principle for human development, growth and aging. *Journal of Theoretical Biology*, *150*, 215–223.

Aoki, I. (1992). Entropy physiology of swine—a macroscopic viewpoint. *Journal of Theoretical Biology*, *157*, 363–371.

Aoki, I. (1994). Entropy production in human life span: A thermodynamical measure for aging. *Age*, *17*, 29–31.

Aoki, I. (1995). Entropy production in living systems: From organisms to ecosystems. *Thermochimica Acta*, *250*, 359–370.

Aoki, I. (1998). Entropy and exergy in the development of living systems: A case study of lake-ecosystems. *Journal of the Physical Society of Japan*, *67*, 2132–2139.

Aoki, I. (2003). Diversity-productivity-stability relationship in freshwater ecosystems: Whole-systemic view of all trophic levels. *Ecological Research*, *18*, 397–404.

Aoki, I. (2006). Min–Max principle of entropy production with time in aquatic communities. *Ecological Complexity*, *3*, 56–63.

Aoki, I. (2008). Entropy law in aquatic communities and general entropy principle for the development of living systems. *Ecological Modelling*, *215*, 89–92.

Aoki, I., & Mizushima, T. (2001). Biomass diversity and stability of food webs in aquatic ecosystems. *Ecological Research*, *16*, 65–71.

Aravindan, C. M. (1993). Preliminary trophic model of Veli Lake, Southern India. In V. Christensen & D. Pauly (Eds.), *Trophic models of aquatic ecosystems, ICLARM conference proceedings* (Vol. 26, pp. 87–89). Manila: International Center for Living Aquatic Resources Management.

Aruga, Y. (1973). *Production in aquatic plant communities II*. Tokyo: Kyoritsu Shuppan (in Japanese).

Barica, J. (1993). Ecosystem stability and sustainability: A lesson from algae. *Verhandlungen Internationale Vereinigung Limnologie*, *25*, 307–311.

Barr, D. P., & DuBois, E. F. (1918). The metabolism in malarial fever. *Archives of Internal Medicine*, *21*, 627–658.

Bartlett, P. N., & Gates, D. M. (1967). The energy budget of a lizard on a tree trunk. *Ecology*, *48*, 315–322.

Battan, L. J. (1979). *Fundamentals of meteorology*. Englewood Cliffs, NJ: Prentice-Hall.

Beattie, J. A., & Oppenheim, I. (1979). *Principles of thermodynamics*. Amsterdam: Elsevier.

Begon, M., Harper, J. L., & Townsend, C. R. (1996). *Ecology* (3rd ed.). Oxford: Blackwell Science Ltd.

Beier, W. (1962). *Biophysik*. VER Georg Thieme.

Bennett, A. F. (1983). Ecological consequences of activity metabolism. In R. B. Huey, E. R. Pianka, & T. W. Schoener (Eds.), *Lizard ecology* (pp. 11–23). Cambridge, MA: Harvard University Press.

Bennett, C. H. (1987). Demons, engines and the second law. *Scientific American*, *257*, 108–116.

Berry, S. (1995). Entropy, irreversibility and evolution. *Journal of Theoretical Biology*, *175*, 197–202.

Birge, E. A. (1915). The heat budgets of American and European lakes. *Transactions of the Wisconsin Academy of Sciences, Arts and Letters*, *18*, 166–213.

Bond, T. E., Kelly, C. F., & Heitman, H., Jr. (1952). Heat and moisture loss from swine. *Agricultural Engineering*, *33*, 148–154.

Brock, T. D. (1985). *A eutrophic lake, Lake Mendota, Wisconsin*. New York, NY: Springer.

Brody, S., & Kibler, H. H. (1944). Resting energy metabolism and pulmonary ventilation in growing swine. *University of Missouri Agricultural Experiment Station Research Bulletin*, *380*, 1–20.

Browder, J. A. (1993). A pilot model of the Gulf of Mexico continental shelf. In V. Christensen & D. Pauly (Eds.), *Trophic models of aquatic ecosystems, ICLARM conference proceedings* (Vol. 26, pp. 279–284). Manila: International Center for Living Aquatic Resources Management.

Budyko, M. I. (1974). *Climate and life*. New York, NY: Academic Press.

Butchbaker, A. F., & Shanklin, M. D. (1964). Partitional heat losses of newborn pigs as affected by air temperature, absolute humidity, age and body weight. *Transactions of the American Society of Agricultural Engineers*, *7*, 380–387.

Christensen, V. (1995). Ecosystem maturity—towards quantification. *Ecological Modelling*, *77*, 3–32.

Christensen, V., & Pauly, D. (1992). ECOPATH II—a software for balancing steady-state ecosystem models and calculating network characteristics. *Ecological Modelling*, *61*, 169–185.

Christensen, V., & Pauly, D. (Eds.), (1993). *Trophic models of aquatic ecosystems, ICLARM conference proceedings* (Vol. 26, p. 390). Manila: International Center for Living Aquatic Resources Management.

Clausius, R. (1865). Über vershiedene für die Anwendungen bequeme Formen der Hauptgleichungen der mechanichen Wärmetheorie. *Poggendorffs Ann. d. Phys*, *125*, 353–400.

Collins, W., Colman, R., Haywood, J., Manning, M. R., & Mote, P. (2007). The physical science behind climate change. *Scientific American*, *297*, 48–55.

Degnbol, P. (1993). The pelagic zone of central Lake Malawi—a trophic box model. In V. Christensen & D. Pauly (Eds.), *Trophic models of aquatic ecosystems, ICLARM conference proceedings* (Vol. 26, pp. 110–115). Manila: International Center for Living Aquatic Resources Management.

Deighton, T. (1932). The determination of the surface area of swine and other animals. *Journal of Agricultural Science*, *22*, 418–449.

DuBois, E. F. (1939). Heat loss from the human body. *Bulletin of New York Academy of Medicine*, *15*, 143–173.

DuBois, E. F., Ebaugh, F. G., Jr., & Hardy, J. D. (1952). Basal heat production and elimination of thirteen normal women at temperatures from 22°C to 35°C. *Journal of Nutrition*, *48*, 257–293.

Dutton, J. A., & Bryson, R. A. (1962). Heat flux in Lake Mendota. *Limnology and Oceanography*, *7*, 80–97.

Gates, D. M. (1963). Leaf temperature and energy exchange. *Archiv für Meteorologie, Geophysik und Bioklimatologie Series B*, *12*, 321–336.

Gates, D. M. (1964). Leaf temperature and transpiration. *Agronomy Journal*, *56*, 273–277.

Gates, D. M. (1980). *Biophysical ecology*. New York, NY: Springer.

Gates, D. M., Tibbals, E. C., & Kreith, F. (1965). Radiation and convection for ponderosa pine. *American Journal of Botany*, *52*, 66–71.

Haase, R. (1969). *Thermodynamics of irreversible processes*. Reading, MA: Addison-Wesly.

Hardy, J. D. (1934). The radiation of heat from the human body III. The human skin as a black-body radiator. *The Journal of Clinical Investigation*, *13*, 615–620.

Hardy, J. D., & DuBois, E. F. (1938a). The technique of measuring radiation and convection. *Journal of Nutrition*, *15*, 461–475.

Hardy, J. D., & DuBois, E. F. (1938b). Basal metabolism, radiation, convection and vaporization at temperatures of 22 to 35°C. *Journal of Nutrition*, *15*, 477–497.

Hardy, J. D., Milhorat, A. T., & DuBois, E. F. (1938a). The effect of forced air currents and clothing on radiation and convection. *Journal of Nutrition*, *15*, 583–595.

Hardy, J. D., Milhorat, A. T., & DuBois, E. F. (1938b). The effect of exercise and chills on heat loss from the nude body. *Journal of Nutrition, 16*, 477–492.

Heymans, J. J., & Baird, D. (2000). Network analysis of the northern Benguela ecosystem by means of NETWRK and ECOPATH. *Ecological Modelling, 131*, 97–119.

Hogetsu, K., Kitazawa, Y., Kurasawa, H., Shiraishi, Y., & Ichimura, S. (1952). Basic study of production and circulation of matter in inland water surface. *Suisankenkyu-kaiho, 4*, 41–127 (in Japanese).

Humphreys, W. F. (1979). Production and respiration in animal populations. *Journal of Animal Ecology, 48*, 427–453.

Hutchinson, G. E. (1957). *A treatise on limnology, I.* New York, NY: Wiley.

Hutchinson, G. E. (1964). The lacustrine microcosm reconsidered. *American Scientist, 52*, 334–341.

Ikusima, I. (1966). Ecological studies on the productivity of aquatic plant communities, II. Seasonal changes in standing crop and productivity of a natural submerged community of *Vallisneria denseserrulata. Botanical Magazine Tokyo, 79*, 7–19.

International Biological Programme Publications Committee. (1967–1975). *IBP Handbook*, Nos. 1–24. Blackwell Scientific Publishers.

International Lake Environment Committee (1994). *Data book of world lake environments* (Vols. 1–3). Kusatsu: International Lake Environment Committee Foundation (ILEC).

IPCC Report of WGI. (2007). <http:// www.ipcc.ch> Accessed 26.03.08.

Ito, K., & Okamoto, I. (1974). Time variation of water temperature in Lake Biwa-ko (VIII). *Japanese Journal of Limnology, 35*, 127–135 (in Japanese with English summary).

Jørgensen, S. E. (2002). *Integration of ecosystem theories: A pattern* (3rd ed.). Dordrecht: Kluwer Academic.

Kauffman, S. (1995). *At home in the universe.* Oxford: Oxford University Press.

Kauffman, S. (2000). *Investigations.* Oxford: Oxford University Press.

Kemp, R. B. (Ed.), (1999). Handbook of thermal analysis and calorimetry (Vol. 4), From Macromolecules to Man. Amsterdam: Elsevier.

Kleiber, M. (1975). *The fire of life.* Malabar, FL: Robert E. Kriegar.

Klekowski, R. Z., & Duncan, A. (1975). Physiological approach to ecological energetics. In W. Grodzinski, R. Z. Klekowski, & A. Duncan (Eds.), *Methods for ecological bioenergetics.* Oxford: Blackwell Scientific Publishers.

Kolding, J. (1993). Trophic interrelationships and community structure at two different periods of Lake Turkana, Kenya: A comparison using the ECOPATH II Box Model. In V. Christensen & D. Pauly (Eds.), *Trophic models of aquatic ecosystems, ICLARM conference proceedings* (Vol. 26, pp. 116–123). Manila: International Center for Living Aquatic Resources Management.

Kotoda, K. (1977). A method for estimating evaporation based on climatological data. *Bulletin of Environmental Research Center University of Tsukuba, 1*, 53–66 (in Japanese).

Kurasawa, H., Tezuka, Y., Kobori, K., & Aoyama, K. (1962). The productions of plankton and large rooted aquatic plants in Usui area of Lake Inba-numa (I). *Miscellaneous Reports of the Research Institute for Natural Resources, 58/59*, 21–36 (in Japanese).

Lampert, W. (1984). The measurement of respiration. In J. A. Downing, & F. H. Rigler (Eds.), *A manual on methods for the assessment of secondary productivity in fresh waters* (2nd ed., pp. 413–468). Blackwell Scientific Publishers.

Lampert, W., & Sommer, U. (1997). *Limnoecology.* Oxford: Oxford University Press.

Lamprecht, I. (2003). Calorimetry and thermodynamics of living systems. *Thermochimica Acta, 405*, 1–13.

Landsberg, P. T. (1961). *Thermodynamics with quantum statistical illustrations*. New York, NY: Interscience.

Landsberg, P. T. (1972). The fourth law of thermodynamics. *Nature*, *238*, 229–231.

Landsberg, P. T., & Tonge, G. (1979). Thermodynamics of the conversion of diluted radiation. *Journal of Physics*, *A12*, 551–562.

Lewis, Christopher J. T. (2007). *Heat and thermodynamics—a historical perspective*. Westport, CT: Greenwood Press.

Lindeman, R. L. (1942). The trophic-dynamic aspect of ecology. *Ecology*, *23*, 399–418.

Liou, K. N. (1980). *An introduction to atmospheric radiation*. New York, NY: Academic Press.

Lovelock, J. E. (1979). *Gaia: A new look at life on earth*. Oxford: Oxford University Press.

Lovelock, J. E. (1988). *The age of gaia*. New York, NY: W. W. Norton.

Maxwell, J. C. (1871). *Theory of heat*. London: Longmans-Green.

Mitamura, O., & Saijo, Y. (1981). Studies on the seasonal changes of dissolved organic carbon, nitrogen, phosphorus and urea concentrations in Lake Biwa. *Archives of Hydrobiology*, *91*, 1–14.

Moen, A. N. (1966). *Factors affecting the energy exchange and movements of white-tailed deer, Western Minnesota*. Ph.D. Thesis, University of Minnesota, St. Paul, MN.

Moen, A. N. (1968a). Energy exchange of white-tailed deer, Western Minnesota. *Ecology*, *49*, 676–682.

Moen, A. N. (1968b). Energy balance of white-tailed deer in the winter. *Transactions of the North American Wildlife and Natural Resources Conference*, *33*, 224–236.

Moen, A. N. (1973). *Wildlife ecology*. New York, NY: W. H. Freeman.

Moen, A. N., & Evans, K. E. (1971). The distribution of energy in relation to snow cover in wildlife habitat. In A. O. Haugen (Ed.), *Proceedings of the symposium on the snow and ice in relation to wildlife and recreation* (pp. 147–162). Ames, IA: Iowa State University.

Monaco, M. E., & Ulanowicz, R. E. (1997). Comparative ecosystem trophic structure of three U. S. mid-Atlantic estuaries. *Marine Ecology Progress Series*, *161*, 239–254.

Monteith, J. L. (1973). *Principles of environmental physics*. London: Edward Arnold.

Moreau, J., Christensen, V., & Pauly, D. (1993). A trophic ecosystem model of Lake George, Uganda. In V. Christensen & D. Pauly (Eds.), *Trophic models of aquatic ecosystems, ICLARM conference proceedings* (Vol. 26, pp. 124–129). Manila: International Center for Living Aquatic Resources Management.

Moreau, J., Ligtvoet, W., & Palomares, M. L. D. (1993). Trophic relationship in the fish community of Lake Victoria, Kenya, with emphasis on the impact of Nile perch (*Lates niloticus*). In V. Christensen & D. Pauly (Eds.), *Trophic models of aquatic ecosystems, ICLARM conference proceedings* (Vol. 26, pp. 144–152). Manila: International Center for Living Aquatic Resources Management.

Moreau, J., Nyakageni, B., Pearce, M., & Petit, P. (1993). Trophic relationships in the pelagic zone of Lake Tanganyika (Burundi Sector). In V. Christensen & D. Pauly (Eds.), *Trophic models of aquatic ecosystems, ICLARM conference proceedings* (Vol. 26, pp. 138–143). Manila: International Center for Living Aquatic Resources Management.

Mori, S., & Yamamoto, G. (Eds.), (1975). *Productivity of communities in Japanese inland waters. JIBP synthesis* (Vol. 10). University of Tokyo Press, 436 pp.

Mount, L. E. (1959). The metabolic rate of the new-born pig in relation to environmental temperature and to age. *Journal of Physiology*, *147*, 333–345.

Mount, L. E. (1964). Radiant and convective heat loss from the new-born pig. *Journal of Physiology*, *173*, 96–113.

Mount, L. E., & Rowell, J. G. (1960a). Body-weight and age in relation to the metabolic-rate of the young pig. *Nature, 186*, 1054–1055.

Mount, L. E., & Rowell, J. G. (1960b). Body size, body temperature and age in relation to the metabolic rate of the pig in the first five weeks after birth. *Journal of Physiology, 154*, 408–416.

Müller, I. (2007). *A history of thermodynamics*. New York, NY: Springer.

Nagi, K. A. (1983). Ecological energetics. In R. B. Huey, E. R. Pianka, & T. W. Schoener (Eds.), *Lizard ecology* (pp. 24–54). Cambridge, MA: Harvard University Press.

National Institute for Research Advancement (1984). *Data book of world lakes*. Tokyo: National Institute for Research Advancement.

Nicolis, G., & Prigogine, I. (1977). *Self-organization in nonequilibrium systems*. New York, NY: Wiley-Interscience.

Nobel, P. S. (1970). *Introduction to biophysical plant physiology*. New York, NY: W. H. Freeman.

Odum, H. T. (1957). Trophic structure and productivity of Silver Springs, Florida. *Ecology Monographs, 27*, 55–112.

Phillipson, J. (1975). Introduction to ecological energetics. In W. Grodzinski, R. Z. Klekowski, & A. Duncan (Eds.), *Methods for ecological bioenergetics*. Cambridge, MA: Blackwell Scientific Publishers.

Planck, M. (1959, 1988). *The theory of heat radiation (Dover)*. Melville, NY: American Institute of Physics.

Prigogine, I. (1967). *Introduction to thermodynamics of irreversible processes* (3rd ed.). New York, NY: Wiley.

Prigogine, I., & Wiame, J. M. (1946). Biologie et Thermodynamique des Phénomènes Irréversibles. *Experientia, 2*, 451–453.

Riley, G. A. (1946). Factors controlling phytoplankton populations on Georges Bank. *Journal of Marine Research, 6*, 54–73.

Rubí, J. M. (2008). The long arm of the second law. *Scientific American, 299*, 40–45.

Saito, N. (2002). Time, to huzio nakano-san and yohei takeda-san. *Butsuri (Physics), 57*, 773–774 (in Japanese).

Sakamoto, M. (1975). Trophic relation and metabolism in ecosystem. In S. Mori & G. Yamamoto (Eds.), *Productivity of communities in Japanese inland waters (JIBP synthesis, Vol. 10)* (pp. 405–410). Tokyo: University of Tokyo Press.

Sasaki, T. (1979). Transition of basal metabolism in Japanese. *Taisha, 16*, 3–12 (in Japanese).

Sasaki, T. (1985). Energy metabolism. In A. Nakayama (Ed.), *Onnetsu seirigaku (thermal physiology)* (pp. 73–95). Rikogakusha (in Japanese).

Schrödinger, E. (1944). *What is life?* Cambridge, MA: Cambridge University Press.

Schwintzer, C. R. (1971). Energy budgets and temperatures of nyctinastic leaves on freezing nights. *Plant Physiology, 48*, 203–207.

Sellers, W. D. (1965). *Physical climatology*. Chicago, IL: The University of Chicago Press.

Serreli, V., Lee, C.-F., Kay, E. R., & Leigh, D. A. (2007). A molecular information ratchet. *Nature, 445*, 523–527.

Shannon, C. E. (1948). A mathematical theory of communication. *The Bell System Technical Journal, 27*, 376–423 , 623–657.

Shelhamer, M. (2007). *Nonlinear dynamics in physiology*. River Edge, NJ: World Scientific.

Shock, N. W. (1955). Metabolism and age. *Journal of Chronic Diseases, 2*, 687–703.

Simpson, E. H. (1949). Measurement of diversity. *Nature, 163*, 688.

Spanner, D. C. (1964). *Introduction of thermodynamics*. New York, NY: Academic Press.

Steemann Nielsen, E. (1960). Productivity of the oceans. *Annual Review of Plant Physiology*, *11*, 341–362.

Stewart, K. M. (1973). Detailed time variations in mean temperature and heat content of some Madison lakes. *Limnology and Oceanography*, *18*, 218–226.

Tanimizu, K., & Miura, T. (1976). Studies on the submerged plant community in Lake Biwa. I. Distribution and productivity of *Egeria densa*, a submerged plant invader, in the south basin. *Physiology and Ecology of Japan*, *17*, 283–290 (in Japanese).

Trincher, K. S. (1967). *Biologie und information*. Teubner.

U. S. Committee for the Global Atmospheric Research Program (1975). *Understanding climatic change*. Washington, DC: National Academy of Sciences.

Walline, P. D., Pisanty, S., Gophen, M., & Berman, T. (1993). The ecosystem of Lake Kinneret, Israel. In V. Christensen & D. Pauly (Eds.), *Trophic models of aquatic ecosystems, ICLARM conference proceedings* (Vol. 26, pp. 103–109). Manila: International Center for Living Aquatic Resources Management.

Wiegert, R. G. (1968). Thermodynamic considerations in animal nutrition. *American Zoologist*, *8*, 71–81.

Zotin, A. I. (1978). The second law, negentropy, thermodynamics of linear irreversible processes. In I. Lamprecht & A. I. Zotin (Eds.), *Thermodynamics of biological processes*. Berlin: Walter de Gruyter.

Zotin, A. I. (1990). *Thermodynamic bases of biological processes*. Berlin: Water de Gruyter.

Zotin, A. I., & Zotina, R. S. (1993). *Phenomenological theory of development, growth and aging*. Hauka (in Russian, with concluding remarks in English).

Printed in the United States
By Bookmasters